DWARF PLANETS AND ASTEROIDS:

MINOR BODIES OF THE SOLAR SYSTEM

By Thomas Wm. Hamilton

Strategic Book Publishing and Rights Co.

Strategic Book Publishing and Rights Co.
12620 FM 1960, Suite A4-507
Houston, TX 77065
www.sbpra.com

ISBN: 978-1-62857-728-0

Book Design: Suzanne Kelly

NAMES: Asteroids named for a person will be noted with the person's dates, and any other associations they may have with asteroids. Ordinary asteroids' names are printed in italics on their first appearance. Dwarf planet names are in bold.

SEARCHES: Several deliberate searches have been made for asteroids. The first was organized around the year 1800 by German astronomers seeking a presumed planet between Mars and Jupiter. They succeeded in finding the second, third, and fourth asteroids. More recent efforts include the Deep Ecliptic Survey to seek small objects beyond Neptune, which was run by the National Optical Astronomy Observatory; LINEAR (the Lincoln Near Earth Asteroid Research) run by the United States Air Force, NASA, and MIT's Lincoln Laboratory; NEAT (Near Earth Astronomical Tracking), operated at Mount Palomar by NASA and Jet Propulsion Lab from December 1995 to April 2007; the Catalina Sky Survey with Siding Spring Survey, operating under a Congressional mandate of 1998 to find ninety percent of potentially hazardous asteroids or comets within ten years; and LONEOS, the Lowell Observatory Near Earth Object Search.

These searches used a variety of instruments. LINEAR, for example, used two one-meter telescopes and a 0.5 meter telescope, and found over 240,000 objects. Catalina/Siding Spring has sixty inch and twenty-seven inch telescopes near Tucson and a twenty inch telescope at Siding Spring in Australia.

DEFINITION

The term *asteroid* was invented by William Herschel (1738-1822; asteroid 2000 Herschel) in 1802. It referred to the star-like appearance of the recently discovered objects. Others have used planetoid or have fudged the issue by calling Ceres a dwarf planet. In this book, asteroid will refer to any object in orbit around the Sun that is less than 1500 kilometers in diameter, more than 0.1 kilometer across, and not a comet. Smaller objects are meteors and not included here. Moons go around planets, and are in a different book, *Moons of the Solar System*, 2013.

DISCOVERY

The existence of asteroids was first suspected by astronomers starting in 1596 with Johannes Kepler (1571-1630), who noted that there seemed to be a large gap among the planets between Mars and Jupiter. In the eighteenth century, German astronomers Johann David Titius (1729-1796) and Johann Elert Bode (1747-1826) noted that each planet going outward from the Sun, seemed to double the distance from the Sun of its predecessor, except between Mars and Jupiter, where the gap more than tripled. Following up on this in 1781 after the discovery of Uranus by William Herschel at what both Titius (asteroid 1998 Titius) and Bode (asteroid 998 Bodea) said was the right distance beyond Saturn, a group of German astronomers organized a committee to search the region between Mars and Jupiter for a dim planet.

While the Germans were still organizing, the Director of Palermo Observatory in Sicily, Giuseppe Piazzi (1746-1826, asteroid 1000 Piazzia), discovered the first asteroid, *Ceres*, on January 1, 1801. Piazzi was checking the accuracy of a recently received catalog and noted an unlisted item, which he soon realized was moving, and, like Herschel with Uranus, initially reported it to be a comet, although, unlike Herschel, he suspected it was something *better*.

Over the next four years following the discovery of Ceres, three more asteroids were discovered, all by the Germans, who had initiated the search. These asteroids were given the names of Pallas, Juno, and Vesta, and, to this day, Ceres, Pallas, and Vesta remain the three largest known asteroids (although Ceres is now classified as a dwarf planet and Vesta's status is disputed).

No further asteroids were discovered until 1845, which inspired new searches, and from 1847 onward, more asteroids were found every year. By the end of the nineteenth century, over 450 asteroids had been found. An astronomer, uninterested in them, described their frequent appearance on his photographs as (translated from German) a plague of minor planets. By the end of the twentieth century, over 7,000 had been catalogued.

DESIGNATIONS AND NAMING

Piazzi chose the name of Ceres for his discovery. Ceres was the patron goddess of Sicily, an agricultural goddess from whom we derive the word *cereal*. The next few asteroids were all named for similar Graeco-Roman mythological females (e.g. Vesta), establishing the rule that asteroids have feminine names. This rule resulted in later asteroids named for countries getting names such as Germania. When asteroids' names began to honor astronomers and other real people, those names would be feminized, such as eminent astronomer Jacobus Kapteyn (1851-1922), whose name was feminized to Kapteynia, as was Piazzi's own asteroid, 1000 Piazzia. After the 1850s, asteroids were given a number also, which was supposed to be sequential in order of discovery. Thus, Ceres became *1 Ceres*, while Kapteyn's became *818 Kapteynia*.

The first asteroid not to have a feminine name was 433 Eros. In addition, Eros was the first discovered asteroid to have an orbit that was not completely between Mars and Jupiter, which created a new custom of giving asteroids with unusual orbits a masculine name. The entire practice of worrying about the gender of asteroidal names was dropped after World War II. Today, an asteroid named for a woman, regardless of location, has a feminine name, and asteroids named for a man have a masculine name, regardless of orbital characteristics.

The first couple thousand asteroids all received names given either by their discoverer, the person who worked out their orbit, or someone given the right by the discoverer (one American astronomer was rumored to sell some naming rights). At some observatories, the directors demanded naming rights, regardless of who at the observatory made the discovery. As astronomers began to honor friends and colleagues by naming an asteroid for them, someone commented that the sky was being turned into a cemetery for astronomers.

Gradually, more people began to receive naming honors, including musicians, philosophers, athletes, and other scientists. Rules were imposed limiting asteroids to one word names of not more than sixteen letters. Those with reasonably unusual names found their last names used with the number; thus, 2308 Schilt for astronomer Jan Schilt of Columbia University (a former student of Kapteyn). Anyone with a fairly common last name would find some version of their first name melded into the last name as a single word, e.g., 4897 Tomhamilton (a former student of Schilt). Very few asteroids slipped through before these rules were imposed.

Following the discovery of a new asteroid, even before it gets a number (which under the rules requires knowing its orbit), it receives a designation that combines the year of discovery with letters and numbers. Thus, Asteroid 2012DA15 indicates that the asteroid was discovered in the year 2012, during the period from February 16 to the end of the month, and was the fifteenth asteroid discovered during that two-week period. Each month is divided in two, with the first fifteen days getting a certain letter (A in January, C in February, E in March, etc.) while the remainder of the month gets the letter that follows (B for January 16 to 31, D for February 16 to 28 or 29, F for March 16 to 31, etc.). The asteroids discovered in the time period are given letters A through Z, and if more are found, numbers are used. (The letter J is never used.)

ORBITS

Any orbit can be defined numerically by six characteristics of the orbit. Certain values of three of these tend to be seen only in the orbits of asteroids, so these are the ones we will discuss here, and include many of the asteroids mentioned below.

First is the semimajor axis. One can think of this as being approximately the average distance of an object from the Sun, or as (aphelion + perihelion)/2. Of course, averages can be misleading. Halley's Comet ranges from inside the orbit of Venus (perihelion of about 66 million miles) to outside the orbit of Neptune (over 2 billion miles for the aphelion), giving an average of over a billion miles, but a distance from the Sun that is actually rare for this comet during its 76 year period. With asteroids, the semimajor axis determines the asteroid's status as a member of the Main Belt of asteroids, those between the orbits of Mars and Jupiter, or a member of less frequently seen groups such as Amors, Atens, Apollos, Trojans, centauroids, etc. Each of these will be discussed.

The semimajor axis is directly related to how long an object takes to complete one orbit around the Sun, using Kepler's Third Law, ka^3=P^2. The *k* just tells us to use the correct units (the year and the Astronomical Unit, usually). The *a* is the semimajor axis in Astronomical Units (1 AU = 92,955,807 miles, or 149,597,870 kilometers). The *P* is the orbital period around the Sun in Earth years. All distances in this book are quoted in Astronomical Units, or AU.

The second number is the orbital eccentricity, which defines how much the orbit deviates from a perfect circle. Asteroids tend to have orbits more elliptical than the orbits of planets, but there are plenty of exceptions. An eccentricity of *0* is a perfect circle. An eccentricity of *1* would be a parabola, which would mean that the object is on its way out of the Solar System, never to return. Eccentricities greater than 1 are hyperbolas, leaving the Solar System fast. The eccentricity is based on the ratio between the semimajor axis and the orbit's closest approach to the Sun. The semimajor axis and eccentricity together define the size and shape of an orbit.

For a semimajor axis *a* and eccentricity *e* perihelion (the point nearest the Sun) is at a distance of a(1-e), while aphelion (furthest from the Sun) is at a(1+e). An ellipse, of course, has two focuses (or foci). The Sun is at one focus, the other is empty space and of no particular importance. In this book, we give the aphelion and perihelion for each asteroid in Astronomical Units (AU), which may be converted to miles or kilometers if readers wish. The eccentricity is also given, and is always less than 1.

The third number is the inclination, the tilt of the orbit with respect to the Earth's orbit (technically known as the ecliptic). Asteroids tend to have orbits inclined to the ecliptic more than orbits of planets, but there are plenty of exceptions.

TYPES

Asteroids are divided into types according to their orbits and according to their geological makeup. The latter is obviously a lot harder to determine than an orbit, except that meteorites are believed to originate from the Main Belt of asteroids, and modern spectroscopy can determine some of the mineralogy. Thus, one of the reasons for studying meteorites is to learn more about the composition of asteroids.

Orbital Types

Asteroids whose orbits are completely contained between the orbits of Mars and Jupiter are called Main Belt asteroids. All the asteroids discovered in the nineteenth century were Main Belt, but since then many others have been discovered that fall into a number of different groups. The belt itself is subdivided into inner and outer belts.

Daniel Kirkwood (1814-1895; asteroid 1578 Kirkwood), an American astronomer teaching at Indiana University, discovered in 1857 that when he plotted the semimajor axis of the asteroid orbits then known, there were gaps or distances (and hence, by Kepler's third law, orbital periods) that seemed to be avoided. He quickly determined that these gaps corresponded to orbital periods that were simple fractions of Jupiter's orbital period. Jupiter takes 11.88 years to orbit the Sun, and the gaps are for asteroids' orbital periods avoiding half, one third, two thirds, one quarter, etc. of 11.88 years. Kirkwood explained this by noting that asteroids in such orbits would periodically receive a kick from Jupiter's gravity which would move them into a different orbit.

It should not be assumed that asteroids are never found at these distances. Their orbits are so eccentric that many will pass through the gap. It is just that no asteroid will have a semimajor axis matching the gaps.

Asteroids in the following categories will be regarded in this book as falling into groups. Thus, the first group described immediately below is the Trojan group. Members of groups have no particular physical relationship; but have been captured into similar orbits, mostly by Jupiter.

Trojan asteroids have orbits they share with a planet. Jupiter has the most such asteroids, but Earth, Mars, Saturn, Uranus, and Neptune are all known to have one or more Trojans. The origin of the collective name derives from the practice, when Jupiter's asteroids were first discovered, of naming them for characters in Homer's work on the Trojan War, *The Illiad*. Trojan asteroids orbit around one of the Lagrangian points. In 1772, Joseph Louis Lagrange (1736-1813) published his finding that there are five points in space where an object of trivial mass could orbit in the same time as a much more massive object orbited with a third, very massive object. An example would be the Sun and Jupiter for the more massive objects, compared to which any asteroid would be trivial. Three of the points Lagrange discovered lie along a straight line.

Let S stand for the Sun, and J for Jupiter, the three points would be

L2-J--L1---------S------------L3. Unfortunately, on a long-term basis, these points are unstable due to the gravitational influence of other planets, and anything at them would ultimately drift away.

The remaining points of L4 and L5, however, would share the orbit of Jupiter, traveling sixty degrees ahead and sixty degrees behind the planet. Not only are these stable on a long-term basis, but objects do not have to be precisely in place; they can, in fact, orbit around the L4 and L5 positions. L4 leads Jupiter by sixty degrees and is called the Greek position, although, unfortunately, some of Troy's warriors have namesakes there. L5 is the Trojan group, but again, a few Greeks have infiltrated. Sixty degrees means that Jupiter, the Sun, and these asteroids form an equilateral triangle. Since the asteroids can orbit around the L4 and L5 locations, there are actually hundreds of asteroids which seem to have drifted into such a position and been captured. Jupiter has 6,035 known Trojans, with L4 having about two thirds of them. Neptune has nine Trojans, Earth and Uranus each have one, Mars two.

- *Amor* asteroids have their aphelion in the Main Belt between Mars and Jupiter, but their orbits cross that of Mars, so their perihelion is closer to the Sun than is Mars, but further from it than Earth. They are named for an asteroid discovered in 1932. Thus far, 4,119 Amor asteroids have been discovered.

- *Apollo* asteroids cross the orbits of both Mars and Earth, so they have an aphelion greater than the orbit of Mars and a perihelion less than the orbit of Earth. They are also named for an asteroid discovered in 1932, not for the Apollo Project. There are 4,892 known Apollo asteroids.
- *Aten* asteroids have orbits that cross the orbit of Earth, but do not go out as far as Mars. They are named for an asteroid (2062) discovered by Eleanor Helin, of whom we shall hear more. There are 758 known Aten asteroids.
- *Atira* asteroids have orbits that are completely inside the orbit of Earth. They are named for an asteroid discovered in 2003 from Socorro, New Mexico. Only nineteen Atira asteroids have been found.
- *Damocloids* have highly eccentric orbits extending well beyond Saturn. They are named for the first one discovered, 5335 Damocles.
- *Hilda* asteroids fall between Mars and Jupiter, but have a 3:2 orbital resonance with Jupiter. Before readers start screaming that this contradicts the remarks about Kirkwood Gaps, permit me to explain. Hilda at aphelion alternately approaches Jupiter's L4 and L5 points. Thus, it never gets close enough to Jupiter to get that gravitational kick to change its orbit, so is trapped in a manner similar to Trojan asteroids. There are 3,774 known Hilda asteroids. They are unrelated to one another except for being gravitationally captured.
- *Centaurs* orbit beyond Jupiter. There have been 460 found. *Plutinos* have orbits similar to Pluto. There are 245 of them. Finally *Trans Neptune Objects* (TNO) account for 906 objects.
- *Main Belt* asteroids now number at least 576,902. These are asteroids whose orbits lie entirely between Mars and Jupiter.

Each of the above-named prototype asteroids is included in the section below discussing selected asteroids.

Geological Types

- *A-type asteroids* have achondrite olivine, are red with a moderate albedo, and are quite rare. They are believed to be from the mantle of a differentiated large body that was broken into pieces in some past collision.
- *B-type asteroids* are fairly rare. They are believed to be rich in carbon, but, unlike the carbonaceous C types, have a slight bluish tint and a higher albedo. It is believed they are composed of anhydrous silicates, hydrated clay, organic polymers, magnetite, and sulfides.
- *C-type asteroids* tend to be exceptionally dark and are believed to have a considerable amount of carbon in their makeup. Few meteorites are found on Earth that are of this type. There are two reasons for this: as meteors heat up on entry into our atmosphere, carbon is quickly converted to carbon dioxide, and those that reach the surface will provide a nice fresh source of food for fungi, bacteria, and the like.
- *D-type asteroids* have very low albedo, similar to the C type, but with a distinctly reddish tint. Their surfaces are believed to include organic-rich silicates, anhydrous silicates, and carbon. They may have water on their interiors. Most class-D asteroids are found in the outer part of the asteroid belt or as Trojans in Jupiter's orbit.
- *E-types* have enstatite (MgSiO3) and feldspar as major surface components. They are rare.
- *G-type asteroids* appear to have clays and mica on their surfaces, suggesting a past presence of water.

- *M-type asteroids* are primarily composed of nickel (Ni) and iron (Fe). The majority of meteorites discovered long after they fell are M types, since metal is easily identified, and many of these respond to magnets. They also rust after being exposed on the ground for long periods.
- *P-type asteroids* seem to combine organic compounds, silicates, carbon, and anhydrous compounds on their surfaces, which have a reddish tint. They are concentrated more than 2.6 AU from the Sun, with many having a semi-major axis around 4.0 AU.
- *Q-type asteroids* are noted for having olivine (generally rare in other types), pyroxene, and metals.
- *S-type asteroids* are composed of silicate materials, i.e. are basically stony. The majority of meteorites seen to fall are stony.

Additional types include stony-iron, at least one asteroid known to be covered in ice, and Vesta, which seems to have a unique surface.

Asteroids sharing geology may or may not be related. In this book, they are referred to as sharing a *class*.

FAMILIES

In 1918, the Japanese astronomer Kiyotsugu Hirayama (1874-1943) noted that there were groups of asteroids sharing similar orbits. He explained these as resulting from collisions among asteroids. The possibility of such collisions should not be taken as evidence that collisions are a frequent occurrence in a crowded asteroid belt. Evidence suggests collisions happen every ten or twenty million years. Identified major families are:

Adeona
Alinda
Dora
Eos
Eunomia
Flora
Gefion
Hansa
Hilda
Hygiea
Koronis
Maria
Nysa
Themis
Vesta

NASA announced on May 30, 2013 that the WISE (Wide-field Infrared Survey Explorer) satellite had identified a total of 76 families of asteroids after studying 120,000 Main Belt objects. There were 38,000 asteroids within these families, of which 26 families were previously unknown.

Descriptions of families appear immediately after the description of their namesake asteroid in this book. Such descriptions are highlighted by the symbol # in the left column.

COLLECTIVES

Brian Marsden (1937-2010), who was chair of the IAU's commission for naming small bodies (asteroid 1877 Marsden), told me that the following relationships among named asteroids was unique in its length, and that was before several of the European members were added.

Jacobus Kapteyn (1851-1922), in addition to having a lunar crater and a red dwarf named for him, has asteroid 818 Kapteynia. At least nine of Kapteyn's former students went on to have asteroids named for them. These include Jean-Jacques Raimond Jr. (1903-1961), asteroid 1450 Raimonda; Willem de Sitter (1872-1934), asteroid 1686 deSitter (and a lunar crater along with a theory of the universe); Jan Oort (1900-1992) (who also has the Oort Cloud), asteroid 1691 Oort; Jan Woltjer (1891-1946), asteroid 1795 Woltjer; Bart Bok (1906-1983) (shared with his wife Priscilla), asteroid 1983 Bok; Adriaan Blaauw (1914-2010), asteroid 2145 Blaauw; Pieter van Rhijn (1886-1968), asteroid 2203 Vanrhijn; Jan Schilt (1894-1982) asteroid 2308 Schilt; and Elly M. Berkhuijsen, asteroid 3604 Berkhuijsen.

Gerard P. Kuiper (1905-1973), asteroid 1776 Kuiper, was a student of Woltjer, de Sitter, and Oort. Kuiper also has the Kuiper Belt and craters on the Moon, Mercury, and Mars named for him. Maarten Schmidt (b. 1939) (asteroid 10430 Martschmidt) was a student of Oort.

Schilt completed his Ph.D. in 1924, two years after Kapteyn died, so he completed his studies under van Rhijn. Schilt had me as a student, asteroid 4897 Tomhamilton, and I had as students Fred Espenak (b. 1952) asteroid 14120 Espenak; and Sheldon Schafer (b. 1948), asteroid 17601 Sheldonschafer.

Among these asteroids, Kuiper has the least eccentric orbit (.013), while Woltjer has the most (.188). The largest semimajor axis belongs to de Sitter (3.41 AU), while the smallest goes to Sheldon Schafer (1.819 AU). These two thus also have the longest and shortest periods, 5.616 and 2.45 years respectively. Inclinations range from de Sitter's 0.623 degrees to Espenak's 24.204 degrees. Appropriately, for the source of this collective, Kapteyn has the brightest asteroid with a magnitude of +9.10. Dimmest is Espenak's +15.5. Only six have had their diameters measured, and Kapteyn is the largest at 49.95 miles. Only three have figures for rotational period, with Oort's 10.27 hours the fastest and Kapteyn's 16.35 hours the slowest. We might note here that Kapteyn comes from a long line of distinguished academics, including Euler and Kepler.

There may well be a few more members of this group in Europe that I have not traced, but it is already an impressively large collection. None of these asteroids seems to be included in any of the more than seventy asteroid families that are known or suspected, and all are Main Belt. Since the term "asteroid family" is already in use for another purpose, I propose that this group, and any similar ones that may exist, be known as *collectives*, with the first name of the originator. Thus, this would be the Jacobus Collective. For those who like observational themes and have already exhausted the Messier list, this might be a novel list to explore and perhaps even contribute to by determining some rotational periods.

We might add that while not all of Kapteyn's academic predecessors have asteroids, they include a few well-known names, such as Leonhard Euler (asteroid 2002 Euler), Gottfried Leibniz, and Johannes Kepler (asteroid 1134 Kepler).

Another collective can be built around the American astronomer Eugene Shoemaker.

Asteroids Visited By Spacecraft			
Spacecraft	***Sponsor***	***Asteroid***	***Year***
Galileo	NASA	Gaspra	1991
Galileo	NASA	Ida/Dactyl	1993
NEAR-Shoemaker	NASA	Matilda	1997
Deep Space 1	NASA	Braille	1999
NEAR-Shoemaker	NASA	Eros	2001
Stardust	NASA	Annefrank	2002
Hayabusa	JAXA	Itokawa	2005
Rosetta	ESA	Steins	2008
Rosetta	ESA	Lutetia	2010
Dawn	NASA	Vesta	2011
Scheduled			
Dawn	NASA	Ceres	2015
Osiris-Rex	NASA	1999RQ36	2018
Hayabusa 2	JAXA	1999JU3	2019
Deep Impact	NASA	2002GT	2020*
*This spacecraft was announced in September 2013 to have failed, so this flyby will not happen.			

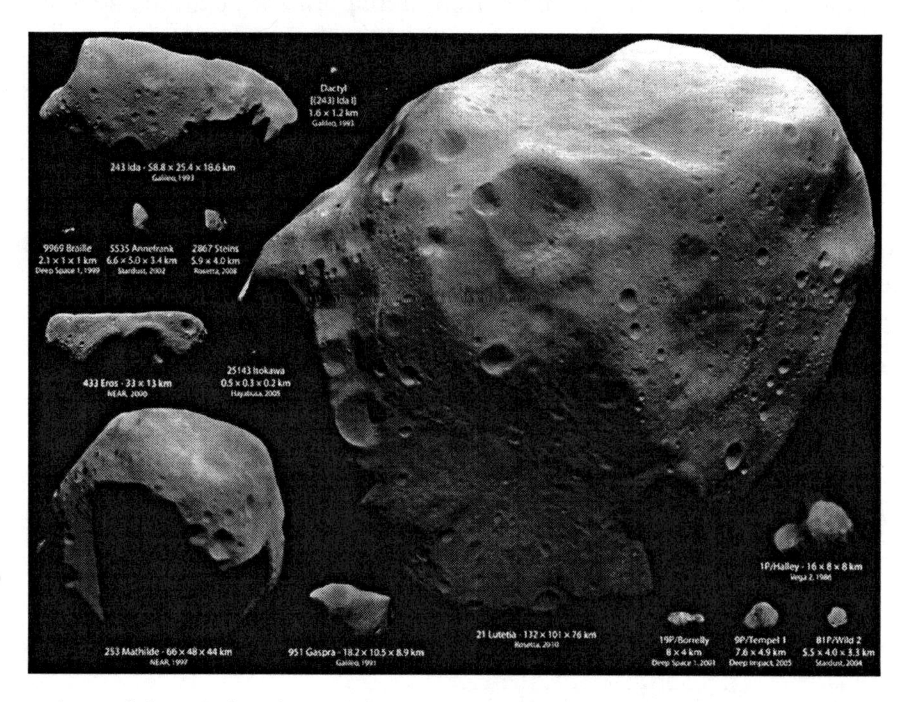

Comparison of the relative sizes of those asteroids that have been visited by spacecraft.

Asteroidal Moons

Over two-hundred asteroids have been discovered to have moons. The first suggestion of this came in the 1970s from David W. Dunham (b. 1932, asteroid 3128 Dunham), who had developed a technique known as stellar occultations to try to determine the size and shape of asteroids. This required a number of volunteers who would be arranged in a line several miles long. This line was set up perpendicular to the expected path where it was hoped an asteroid would briefly block (or occult, in the technical terminology) a star from view as the asteroid moved in its orbit through space. By timing the duration of the occultation at each location, the size and shape of the occulting asteroid could be determined. Dunham was surprised to find that in a few cases the star disappeared, reappeared, and then disappeared and reappeared a second time, or that the star never disappeared for one or more observers, while vanishing for observers on either side. The explanation was that the asteroid had another body, a moon, accompanying it.

Dunham's discovery was immediately controversial, as many doubted that an asteroid would have a strong enough gravity field to hold onto a moon. At the time I noted in a letter to *Sky & Telescope* magazine and to my classes, that pairs of large craters of similar age are seen on the Moon, Mars, and even Earth, where the Clearwater Lakes of Canada provide an outstanding example of a location where an asteroid impacted several hundred million years ago to create a lake twenty miles across, while its moon created a lake thirteen miles across. Today the existence of asteroidal moons is no longer controversial, although their origins still cause disputes.

Asteroidal moons are described immediately after their parent body in this book. Such moons are highlighted by an asterisk (*) in the left column.

Selected Individual Asteroids

1 Ceres was discovered on January 1, 1801, by Giuseppe Piazzi (1746-1826; asteroid 1000 Piazzia) at Palermo Observatory. By February 11 it had vanished from view as it approached solar conjunction. The mathematician Carl Friedrich Gauss (1777-1855; asteroid 1001 Gauss) then invented the method still used today to determine orbits from a limited number of observations (although Gauss had to do this with pen and paper, while today a computer does the hard work), and Ceres was recovered on December 31, 1801. It is the largest known asteroid, with a diameter of 590 miles. This is enough to give it a spherical shape, and Ceres is today sometimes regarded as a dwarf planet. It is believed to have a rocky core with an icy mantle and a surface of carbonates and clays.

The Hubble Space Telescope has photographed Ceres a number of times, revealing surface features that are believed to be craters. The Dawn spacecraft will arrive at Ceres in 2015 and either verify or correct this.

Ceres's orbit has an aphelion of 2.9858 AU, and a perihelion of 2.5468 AU. The eccentricity is 0.079138. The inclination is 10.587 degrees. Ceres takes 1679.67 days, or 4.60 years to orbit the Sun. The escape velocity from the surface is 1,670 feet per second (for comparison, the escape velocity of our Moon is 1.2 miles per second). Ceres rotates in 9 hours 4 minutes.

While the mass will be better measured when Dawn arrives, current calculations give it a mass 4% that of the Moon, yielding a density of 2.077. A high temperature of -36F has been detected, although this surely is an extreme high.

2 Pallas was discovered by Heinrich Wilhelm Olbers (1758-1840, asteroid 1002 Olbersia) on March 28, 1802. It has a diameter of 338 miles. Pallas is an alternative name for the goddess Athena.

Pallas has an aphelion of 3.412 AU, a perihelion of 2.192 AU, and therefore a fairly high orbital eccentricity of 0.231. The inclination is a very high 34.84 degrees, so it is rarely found close to the ecliptic. (This will make sending a spacecraft to visit rather difficult.) The orbital period is 1685.87 days, or 4.62 years. Pallas rotates in 7.8132 hours.

This asteroid is barely capable, on rare occasions, of being visible in a dark clear sky to people with excellent eyesight, as it has an apparent magnitude from 6.49 to 10.68.

The escape velocity is 115 feet per second, thanks to a surface gravity .2% that of Earth. Density is 2.8.

Pallas is classified as having one of the rarer types of geology, a B class asteroid. This class is noted for being rich in volatile carbon compounds. There is one observation claiming to have detected a satellite less than a mile in diameter. This is unconfirmed.

Pallas is the main member of a minor family of about two dozen asteroids, all with similar inclinations, semimajor axis, albedo, and spectra.

3 Juno was discovered by Karl Ludwig Harding (1765-1834, asteroid 2003 Harding) on September 1, 1804. It has a diameter of 145 miles. Juno was the wife of Jupiter.

Juno has an aphelion of 3.352587 AU, and a perihelion of 1.98888 AU, yielding an eccentricity of 0.2553. This is the second highest eccentricity among asteroids with a diameter of more than 120 miles. The inclination is 12.979 degrees. The orbital period is 4.36 years, or 1594.206 days. Juno rotates on its axis in 7.210 hours. Earth based observations suggest Juno may have a crater 60 miles across. This asteroid's shape seems to be similar to a rectangular, well plumped pillow.

The escape velocity is about 62 feet per second, with the surface gravity about 0.13% that of Earth.

The apparent magnitude of Juno ranges from 7.4 to 11.55. It was involved in the first ever observed occultation of a star by an asteroid in 1988.

Juno has an S type geology, with silicate compounds predominating.

4 Vesta was discovered by Heinrich Olbers, the same person who discovered 2 Pallas, on March 29, 1807. It has a diameter of 328 miles. Vesta was a goddess of the hearth and home.

Vesta has an aphelion of 2.5712 AU, and a perihelion of 2.1526 AU, with an eccentricity of 0.08862. The inclination is 7.134 degrees. The orbital period is 3.63 years, or 1325.849 days. The rotational period is 5.342 hours.

The surface gravity is 0.25% that of Earth. Vesta has a remarkably high density, 3.456, one of the few asteroids known to have a density higher than that of our Moon.

Vesta can be visible to the unaided eye in clear dark skies when opposition and perihelion coincide. Its apparent magnitude ranges from 5.1 to 8.48. The cover photograph on this book shows Vesta from the Dawn spacecraft in 2012.

Vesta has an unusual geology, and is apparently differentiated, that is, the denser materials are concentrated in a core of nickel iron 125 to 130 miles in diameter, with olivine [$(MgFe)_2SiO_4$] forming layers of mantle above the core, topped by a crust. Since the Dawn spacecraft spent over a year in orbit around Vesta, its surface was mapped, and mineralogy studied from orbit. The South Polar region has an enormous crater, Rheasylvia, that seems to be about one billion years old. From the size of the gouge, Vesta must have lost about one percent of its mass.

This is believed to be the origin of the Vesta family of asteroids, as well as the source of some unusual types of meteorites that have landed on Earth. The orbital examination of this asteroid seems to

confirm the source of the meteorites. Members of this family have a semimajor axis in the range of 2.26 to 2.48 AU, with an eccentricity between .03 and .16. Orbits are inclined to the ecliptic by 5.0 to 8.3 degrees. About 250 members of this family have been identified.

5 Astraea was discovered on December 8, 1845, by Karl Ludwig Hencke (1793-1866, asteroid 2005 Hencke). It has a diameter of 73.8 miles. Astraea was a goddess of justice.

Astraea has an aphelion of 3.0659 AU, a perihelion of 2.0836 AH, and an eccentricity of .19074. The inclination of the orbit is 5.367 degrees. The orbital period is 1509.05 days, or 4.13 years. The rotational period is 16.80 hours.

The escape velocity is 203 feet per second.

The apparent magnitude of Astraea ranges from 8.74 to 12.89.

Geologically Astraea is a class B asteroid, mostly nickel-iron, magnesium, and iron silicates.

6 Hebe was discovered by Karl Hencke on July 1, 1847. It is somewhat irregularly shaped, with a mean diameter of 115 miles. Hebe was a goddess of youth, and the name for the asteroid was proposed by Gauss.

Hebe has an aphelion of 2.914 AU, a perihelion of 1.937 AU, and an eccentricity of 0.202. The inclination is 14.751 degrees. The orbital period is 3.78 years, or 1379.75 days. The rotational period is 7.2744 hours.

Hebe's surface gravity is 0.8% that of Earth, with a very high density of 3.81.

This asteroid has an apparent magnitude of 7.5 to 11.5, and an albedo of 0.268.

Hebe has an S type geology, with about 60% silicates and 40% nickel-iron. Several large craters on the surface seem to have exposed some of the nickel-iron, and an entire class of meteorites found on Earth is believed to have originated on Hebe.

7 Iris was discovered by John Russell Hind (1823-1895, asteroid 1897 Hind) on August 13, 1847. Iris was the first of ten asteroids Hind discovered, along with several variable stars. A crater on the Moon is also named for him. Iris has a diameter of 124 miles. Iris was the goddess of rainbows (thus Sinus Iridium, the Bay of Rainbows, on the Moon), and an attendant on Juno.

Iris has an aphelion of 2.936 AU, a perihelion of 2.386 AU, and an orbital eccentricity of 0.2305. The inclination of the orbit to the ecliptic is 5.5245 degrees. The orbital period is 3.69 years, or 1346.42 days. The rotational period is 7.139 hours.

The escape velocity is 481.3 feet per second. The density is 3.21.

Iris has an apparent magnitude in the range of 6.7 to 11.1. The albedo is 0.2766.

This is another S class geology asteroid. The shape seems to approximate a regular solid of roughly twelve facets.

8 Flora was the second asteroid discovery of Hind, made October 18, 1847. Flora is 85 by 85 by 70 miles. Flora was the goddess of flowers and mother of the goddess of spring. The name was suggested by John Herschel.

This asteroid has an aphelion of 2.546 AU, a perihelion of 1.858 AU, and an eccentricity of 0.1561. The orbit's inclination is 5.886 degrees. The orbital period is 1193.55 days, or 3.27 years. The rotational period is 12.865 hours.

The surface gravity is 0.005G, and the escape velocity is about 525 feet per second.

Flora's apparent magnitude ranges from 7.9 to 11.6. The albedo is .243.

Flora is the namesake of a family of asteroids. They have a semimajor axis between 2.15 and 2.35 AU, an eccentricity of .03 to .23, and an orbital inclination of 1.5 to 8.0 degrees. About six hundred members of this family have been identified.

9 Metis was discovered April 25, 1848, by Andrew Graham (1815-1908). The moon nearest Jupiter is also named Metis. This asteroid has an irregular shape, with a broad end and a pointed end. The mean diameter is 118 miles. Metis was a Titan and the daughter of Tethys and Oceanus.

Metis has an aphelion of 2.6787 AU, and a perihelion of 2.0932 AU, giving an eccentricity of 0.12296. The inclination is 5.575 degrees. The orbital period is 1346.11 days, or 3.69 years. The rotational period is 5.079 hours.

The surface gravity on this asteroid is 0.008G, while the escape velocity is about 350 feet per second. The density is 4.12.

Metis has an apparent magnitude of 8.1 to 11.83.

Geologically it is classed as an S type, and is believed to be the remnant core of a much larger asteroid smashed in a collision over one billion years ago. It is 30 to 40% olivine, with the remainder mostly nickel-iron.

10 Hygeia was discovered on April 12, 1849, by Annibale De Gasparis (1819-1892, asteroid 4279 DeGasparis), the first of nine that he was to find. The diameter is 252.5 miles. Hygeia was the goddess of health.

The aphelion is 3.5001 AU, the perihelion 2.7728 AU, and the orbital eccentricity is 0.1161. The inclination of the orbit to the ecliptic is 3.8419 degrees. The orbital period is 2029.446 days, or 5.56 years. The rotational period is 27.623 hours.

Hygeia's surface gravity is 0.01G, with an escape velocity of 655 feet per second. The density is 2.08.

Hygeia has an apparent magnitude of 9.0 to 11.97, with an albedo of .0717.

Geologically, this is the first class C asteroid we have come to, being basically carbonaceous. This explains its dimness and low albedo.

Hygeia's family of asteroids has semimajor axes between 3.06 and 3.24 AU, with eccentricities between .09 and .19. Orbital inclinations range from 3.5 to 6.8 degrees. Over one hundred members have been identified.

11 Parthenope was discovered May 11, 1850, by De Gasparis, the second of his nine. The name is one of the sirens in Greek mythology. The diameter is 95.3 miles.

Parthenope's orbit has an aphelion of 2.6955 AU, a perihelion of 2.2105 AU, and eccentricity of .0988. The inclination of the orbit to the ecliptic is 4.6261 degrees. The orbital period is 1403.30 days, or 3.84 years. It rotates in 13.72 hours.

This S class asteroid has an albedo of .1803, with a density of 3.28. The escape velocity is 320 feet per second. The apparent magnitude ranges from 8.7 to 12.2.

12 Victoria was discovered by John Russell Hind on September 13, 1850. It is oblong, 69 by 77 miles. The name was controversial, as it appeared Hind was breaking the tradition of using mythological names to honor the then reigning Queen of England. Hind however insisted the name was taken from the Roman goddess of victory (perhaps with a sly wink?), and the name was finally accepted.

Victoria's aphelion is 2.849 AU, the perihelion is 1.819 AU. The orbital eccentricity is .2213. Inclination is 8.363 degrees. The orbital period is 1302.44 days, or 3.17 years. The rotation is 8.66 hours.

Victoria has a surface gravity of 0.003G, and an escape velocity of 200 feet per second. The density is 2.45.

Victoria has an apparent magnitude ranging from 8.68 to 12.82. The albedo is .177. Geologically it is S class.

13 Egeria was discovered November 2, 1850, by De Gasparis. It is named for a nymph who was the wife of the semi-mythical second king of Rome. It has a diameter of 127.5 miles.

The aphelion is 2.7922 AU, the perihelion is 2.3600 AU, and the eccentricity .0839. The inclination of the orbit to the ecliptic is 16.541 degrees. The orbital period is 1510.21 days, or 4.13 years. The rotation is 7.045 hours.

This class G asteroid has an albedo of .0825. Spectral studies suggest a high water content, perhaps as much as 11% by mass, yet the density is a high 3.46, so there must be some fairly dense metals to accompany the water. The surface gravity is 0.0072G (Earth's surface gravity is G=1.0), with an escape velocity of 370 feet per second.

14 Irene was discovered May 19, 1851, by Hind. It is named for one of the Horae who was responsible for peace. The diameter is 93 miles.

Irene's aphelion is 3.0147 AU, the perihelion is 2.1575 AU, and eccentricity is .1657. The orbital inclination is 9.1189 degrees. The period of revolution around the Sun is 1519.03 days, or 4.16 years. It rotates in 15.03 hours.

This S class asteroid has an albedo of .159. The density is 4.42.

The surface gravity is 0.0065G, with an escape velocity of 300 feet per second. The apparent magnitude ranges from 8.85 to 12.30.

15 Eunomia was discovered on July 29, 1851, by De Gasparis. It is named for one of the Horae in charge of law and order, although there is no record of her ever approving reading people's mail. The diameter is 158.7 miles.

The aphelion is 3.1395 AU, the perihelion is 2.1458 AU, and eccentricity .1880. The orbital inclination is 11.74 degrees. The period of the orbit is 1569.14 days, or 4.30 years. It rotates in 6.083 hours.

This largest known class S asteroid has an albedo of .2094. Various regions on the surface differ in their composition, with areas of pyroxenes and olivine, just olivine, and basalt with pyroxene identified. It is believed this indicates at least partial differentiation.

Eunomia is in a 7:16 resonance with Mars.

The Eunomia family of asteroids has a semimajor axis between 2.53 and 2.72 AU, with eccentricities between .08 and .22. The orbits are inclined from 11.1 to 15.8 degrees to the ecliptic. Nearly 400 have been identified.

16 Psyche was discovered March 17, 1852 by De Gasparis. It is named for one of the popular figures in Greek mythology. The diameter is about 157 miles.

Psyche's aphelion is 3.3223 AU, the perihelion is 2.5237 AU, and eccentricity .1366. The orbit is inclined 3.098 degrees to the ecliptic. The orbital period is 1825.33 days, or 5.00 years. It rotates in 4.196 hours.

With an M spectrum, the albedo is .1203, and the density a very high 6.98. The surface gravity is 0.0055G, and escape velocity 426 feet per second.

17 Thetis was discovered April 17, 1852, by K. Robert Luther (asteroid 1308 Luthera), his first asteroid. It is named for the mother of Achilles. The diameter is 56 miles.

The aphelion of Thetis is 2.8007 AU, the perihelion is 2.1405 AU, and eccentricity .1336. The orbital inclination is 5.589 degrees. The orbital period is 1418.39 days, or 3.88 years. It rotates in 12.27 hours.

The surface is believed to be more than 40% pyroxene, and under 20% olivine. The surface gravity is 0.003G, with the escape velocity 156 feet per second.

18 Melpomene was discovered June 25, 1852, by Hind. It is named for the muse of tragedy. The diameter is 87.4 miles.

Melpomene's aphelion is 2.7978 AU, the perihelion is 1.7948 AU, and eccentricity .2184. Inclination of the orbit is 10.139 degrees. The orbital period is 1271.0 days, or 3.48 years. It rotates in 11.57 hours.

The spectrum is B class, with an albedo of .2225. The surface gravity is 0.0046G, with an escape velocity of 244 feet per second. A moon has been suspected but never confirmed.

19 Fortuna was discovered August 22, 1852 by Hind. It is named for the Roman goddess of good luck. The diameter is 124 miles.

Fortuna's aphelion is 2.8294 AU, the perihelion is 2.0564 AU, and eccentricity .1582. The orbit is inclined 1.573 degrees. The orbital period is 1394.62 days, or 3.82 years. It rotates in 7.443 hours.

With a spectral class G, the albedo is .037, making this one of the darkest of the early discovered asteroids. The density is 2.70. The surface gravity is 0.0071G. Tholins are suspected on the surface.

20 Massalia was discovered by De Gasparis on September 19, 1852, and independently by Jean Chacornac (1823-1873, asteroid 1622) on September 20. The diameter is 90 miles. De Gasparis wanted to name it Olympia, while another prominent astronomer had been campaigning to name asteroids after their discoverers (which would have introduced inconvenient ambiguity then, this being De Gasparis's third and the first of Chacornac's six, and total chaos today). Chacornac's choice, giving it the Greek form of the city of Marseille where his observatory was located, was the name that was accepted. This was the first asteroid name not based in mythology, if we accept Victoria.

Massalia's aphelion is 2.752 AU, the perihelion 2.0651 AU. Orbital eccentricity is 0.1426. Inclination is 0.7082 degree. The orbital period is 1365.28 days, or 3.74 years. The rotational period is 8.098 hours.

Massalia's surface gravity is 0.006G, and the escape velocity is 304.5 feet per second. The density is 3.54.

Apparent magnitude for Massalia ranges from 8.3 to 12.0, with an albedo of .210.

Massalia is classified as S, a silicate object.

It seems to have suffered a major impact at some time in the past, as there are about fifty known, very small asteroids in similar orbits, all S class also. Their semimajor axes are 2.37 to 2.45 AU, with eccentricities of .12 to .21. The orbits are inclined 0.4 to 2.4 degrees.

21 Lutetia was discovered November 15, 1852, by H. Goldschmidt. It is named for the ancient Roman city of Lutetia, now known as Paris. It is 59 by 66 miles.

Lutetia's aphelion is 2.8347 AU, the perihelion 2.0340 AU, and eccentricity .1645. The inclination of the orbit to the ecliptic is 3.0639 degrees. The orbital period is 1337.32 days, or 3.80 years. It rotates in 8.1655 hours.

The albedo is .2212, with a class B spectrum. The density is 3.4, among the highest of any asteroid, with a regolith believed to be up to 1.8 miles thick. The surface shows many craters, with scarps, fractures, and grooves.

22 Kalliope was discovered by Hind on November 16, 1852. It is 133 by 110 by 93 miles. The name is taken from the Muse of epic poetry.

Kalliope's aphelion is 3.2009 AU, with a perihelion of 2.6177 AU. The eccentricity is .10024, with an inclination to the ecliptic of 13.719 degrees. The orbital period is 1812.49 days, or 4.96 years. It rotates in 4.148 hours.

Kalliope has a surface gravity of 0.008G, and an escape velocity of 690 feet per second. Its density is 3.4.

This asteroid is classified as M, having nickel and iron, but mixed with hydrates and silicates. The albedo is .17.

* Kalliope is the lowest numbered asteroid found to have a moon. On August 29, 2001, Jean-Luc Margot (b. 1969; asteroid 9531 Jean-Luc) and Michael E. Brown (b. 1965) found a moon Brown chose to name *Linus* (matching his unsuccessful campaign to give Eris the name Xena). This is the second of five asteroidal moons found by Margot, while Brown is well known for discovering Trans-Neptunian objects, including several dwarf planets.

Linus has a diameter of 17 miles, and orbits 660 miles from Kalliope.

23 Thalia was discovered December 15, 1852, by Hind. It is named for the muse of the theater. The diameter is 66.2 miles.

Thalia's aphelion is 3.2428 AU, the perihelion is 2.0076 AU, and eccentricity .2353. The orbit is inclined 10.115 degrees. The orbital period is 1553.61 days, 4.25 years. It rotates in 12.312 hours.

With spectral class S, the albedo is .2536. The surface gravity is 0.006G, with an escape velocity of 186 feet per second.

24 Themis was discovered April 5, 1853, by De Gasparis. Themis was a Greek goddess of law and order. The diameter is 123 miles.

Themis has an aphelion of 3.5368 AU, a perihelion of 2.7400 AU, and an eccentricity of .1269. The orbital inclination is 0.7524 degree. The orbital period is 2030.80 days, or 5.56 years. It rotates in 5.374 hours.

It has a C spectrum, with an albedo of .067. The density is 2.78. The surface appears to be combination of water ice and organic compounds. The surface gravity is 0.0011G, with an escape velocity of 278 feet per second.

The Themis family has semimajor axes between 3.08 and 3.24 AU. Eccentricities range from .09 to .22. Orbits are inclined to the ecliptic by less than 3 degrees. Over 500 are known.

25 Phocaea was discovered April 6, 1853, by Jean Chacornac (1823-1873; the first of six discovered asteroids; 1622 Chacornac). It has the Greek name of a town in Turkey. The diameter is 46 miles.

The aphelion is 3.0136 AU, the perihelion 1.7860 AU, and the eccentricity .2558. The inclination is 21.592 degrees. The orbital period is 1357.85 days, or 3.72 years. It rotates in 9.934 hours.

The albedo is .2310 for this S class. The density is 2.2, with an escape velocity of 133 feet per second, and a surface gravity of 0.0028G.

30 Urania was discovered on July 22, 1854, the last asteroid discovery by John Russell Hind. The diameter is 68 by 56.5 miles. Urania was the Greek Muse of astronomy. (Didn't know we had our own Muse, did you?)

Urania's orbit has an aphelion of 2.666 AU, a perihelion of 2.0645 AU, and an eccentricity of 0.12716. Inclination is 2.098 degrees. The orbital period is 1365.28 days, or 3.74 years. The rotational period is 13.686 hours.

Urania's surface gravity is 0.003G. The escape velocity is 620 feet per second. The density is 3.92.

The apparent magnitude of Urania ranges from 9.36 to 13.

Urania is classified as an S, silicate dominated asteroid.

35 Leukothea was discovered April 19, 1855, by K. Luther. It is named for a minor sea goddess. The diameter is 64 miles.

Leukothea's aphelion is 3.6694 AU, the perihelion 2.3109 AU, and the eccentricity is .2272. The inclination of the orbit is 7.9355 degrees. The period of the orbit around the Sun is 1888.60 days, or 5.17 years. It rotates in 5.357 hours.

This class C asteroid has an albedo of .0662. The escape velocity is 190 feet per second.

40 Harmonia was discovered on March 31, 1856, by Hermann M. S. Goldschmidt (1802-1866, asteroid 1614 Goldschmidt). It was the fourth of fourteen asteroids discovered by Goldschmidt, setting a mid-century record. It was given the name of a goddess of peace in honor of the end of the Crimean War. Harmonia's diameter is 66 miles.

Harmonia has an aphelion of 2.373 AU, and a perihelion of 2.1602 AU. The eccentricity if 0.0471, and the orbital inclination is 4.2577 degrees. Harmonia revolves around the Sun in 1246.74 days, or 3.41 years. The rotational period is 8.91 hours.

Harmonia has a surface gravity of 0.0028G, and an escape velocity of 186.3 feet per second. Density is 2.

The apparent magnitude of Harmonia ranges from 9.31 to 11.8.

This is yet another S class asteroid.

42 Isis was the first asteroid discovery of Norman Pogson, on May 23, 1856. The name was given by the director of the Cambridge Observatory where Pogson was working and where he later became director. It is the first asteroid name admittedly awarded to honor a real person, Pogson's daughter, Elizabeth Isis Pogson, who was an astronomer of note herself. The asteroid's diameter is 62 miles.

Isis has an aphelion of 2.9851 AU, and a perihelion of 1.8989 AU. The orbital eccentricity is 0.2224, and the inclination is 8.5269 degrees. Isis orbits the Sun in 1393.84 days, or 3.82 years. The rotational period is 13.597 hours.

Isis has a surface gravity of .003G, and an escape velocity of 160 feet per second. The density is 2.38.

The apparent magnitude ranges from 9.18 to 13.50. The albedo is .1712.

Isis is classed as an S asteroid, but its spectrum shows olivine, a mildly unusual mineral for an asteroid.

44 Nysa was discovered May 27, 1857, by Goldschmidt. The name was the mythological birthplace of Dionysius. It is very irregularly shaped, 70 by 38 by 36 miles.

Nysa has an aphelion of 2.7826 AU, and a perihelion of 2.0636 AU. The eccentricity is .14836, with an inclination of 3.0762 degrees. The orbital period is 1377.69 days, or 3.77 years. It rotates in 6.422 hours.

Nysa's surface gravity is 0.002G, and the escape velocity is 124 feet per second. Density is 2.0.

The apparent magnitude ranges from 8.83 to 12.46.

Nysa's orbit makes it the largest E class asteroid inside Jupiter's resonance orbits.

Members of Nysa's family of asteroids have semimajor axes of 2.41 to 2.5 AU, and eccentricities of .12 to .21. Orbits are inclined 1.5 to 4.3 degrees. Nearly 400 are known.

45 Eugenia was discovered June 28, 1857, by Goldschmidt. It was named for the wife of Napoleon III, one of the first to acknowledge being named for a real person. The diameter is 132 miles.

The aphelion is 2.9453 AU, the perihelion is 2.4944 AU, and the eccentricity is .0829. The inclination of the orbit is 6.6034 degrees. The orbital period is 1638.36 days, or 4.49 years. It rotates in 5.7 hours.

This F class asteroid has an albedo of .0398. The density is 1.1, possibly suggesting it is partly rubble.

The surface gravity is 0.006G, with an escape velocity of 235 feet per second.

* Despite the low values above, this was the first asteroid identified as having two moons. The first was discovered November 1, 1998, by William J. Merline and colleagues and given the name Petit Prince. The diameter is 8.5 miles, and it orbits Eugenia in 4.76 days in an orbit inclined to Eugenia's equator by 8 degrees.

* The second moon was discovered in 2004, and so far has no more name than S/2004(45) 1. The diameter is four miles.

50 Virginia was discovered on October 4, 1857 (exactly 100 years before the first Earth satellite launch) by James Ferguson (1797-1867; asteroid 1745 Ferguson), one of the earliest asteroid discoveries made by an American. The diameter is 62 miles, and it is named for the State of Virginia.

Virginia has an aphelion distance of 3.404 AU, perihelion of 1.888 AU, and an eccentricity of 0.2840. The inclination is 2.8333 degrees. However, this asteroid is in an 11:4 orbital resonance with the planet Jupiter, so this orbit is not merely unstable, but chaotic on a time span of 10,000 years. Virginia orbits the Sun in 1576.88 days, or 4.32 years. The rotational period is 14.315 hours.

Surface gravity and escape velocity have not been determined, but are probably similar to those for Harmonia and similar asteroids.

This asteroid has a remarkably high density of 4.49, given that its albedo of .0357 assures it has mostly carbon on its surface.

Virginia is a C class asteroid, carbonaceous, and therefore dim, rarely brighter than tenth magnitude.

60 Echo was the third and last of Ferguson's asteroid discoveries, made September 18, 1860. The diameter is 37 miles. It is named for a nymph in Greek mythology. Just a century later the United States launched into Earth's orbit the first of two 100 foot diameter balloons named Echo 1 and Echo 2. They were used for reflecting and thus relaying radio messages. (There was also a British columnist, Ivan Robinson, who used the pen name Echo 3 for his column.)

Echo has an aphelion of 2.8339 AU, while the perihelion is 1.9527, with eccentricity of the orbit at 0.1841. The inclination is 3.601 degrees. Echo goes around the Sun in 1352.36 days, or 3.70 years. It rotates in 25.208 hours.

The surface gravity on Echo is 0.003G, and the escape velocity is around 200 feet per second.
Density is 2.78, with an albedo of 0.2535.
Echo is an S class asteroid.

65 Cybele was discovered by Ernst W. Tempel (1821-1889; asteroid 3808 Tempel) on March 8, 1861. It was the second of five asteroids he discovered, along with quite a few comets. Cybele was the Earth goddess of the Phrygians and a recognized oracle for the Greeks. The diameter is 192 miles, making it the seventh largest known asteroid.

Cybele has an aphelion of 3.8031 AU, a perihelion of 3.0485 AU, and an eccentricity of .11013. The orbital inclination is 3.5629 degrees. Cybele's orbital period is 2316.03 days, or 6.34 years. It rotates in 6.08 hours.

Surface gravity is 0.0067G, with an escape velocity of 480 feet per second. Density is 1.78, suggesting a rather porous interior. Albedo is .0706, with a surface that seems to be a mixture of water ice and carbonaceous dust.

This is the namesake for the Cybele family of asteroids, characterized by a semimajor axis of 3.27 to 3.7 AU, an eccentricity <.3 and an inclination <25 degrees.

* A stellar occultation suggested the possibility of a seven-mile diameter moon for Cybele.

70 Panopaea is the last of Goldschmidt's asteroid discoveries, found on May 5, 1861. The diameter is 76 miles. It is named for a Greek nymph.

Panopaea has an aphelion of 3.0915 AU, and a perihelion of 2.1383 AU. The orbital eccentricity is 0.18227. The inclination of the orbit to the ecliptic is 11.5915 degrees. Panopaea orbits the Sun in 1544.46 days, or 4.23 years. Because of a resonance with both Jupiter and Saturn, this orbit will be changed chaotically in perhaps 20,000 years. It rotates in 15.797 hours.

The density is 3.48, rather high for a class C asteroid. The albedo is .0675, suggesting the surface may have some slightly lighter colored materials in addition to the expected carbonaceous.

75 Eurydike was discovered September 22, 1862, by Christian H. F. Peters. The name is the German form of Euridice, wife of Orpheus. The diameter is 34.6 miles.

The aphelion is 3.4879 AU, the perihelion I 1.8541 AU, and the eccentricity .3059. The inclination of the orbit is 5.0023 degrees. Eurydike has an orbital period of 1594.43 days, or 4.37 years. It rotates in 5.357 hours.

This class M asteroid has an albedo of .149. The surface gravity is 0.005G, and the escape velocity 100 feet per second.

80 Sappho was discovered on May 2, 1864, by Norman Pogson (1829-1891; asteroid 1830 Pogson), one of eight asteroids he found. The diameter is 48.6 miles. Sappho was a famous woman poet of Ancient Greece, specializing (so far as that which has come down to us would indicate) in love poetry.

Sappho has an aphelion of 2.7554 AU, and a perihelion of 1.8354 AU. The eccentricity is .20041. The inclination is 8.6658 degrees. Sappho goes around the Sun in 1270.25 days, or 3.48 years. It rotates in 14.03 hours.

The surface gravity is 0.0028 G, with an escape velocity of 346 feet per second.
Sappho's albedo is .1848.
This is a class S asteroid.

87 Sylvia was another of Pogson's discoveries, named for the mythological mother of Romulus and Remus. It was discovered on May 16, 1866.

Sylvia has an aphelion of 3.8006 AU and a perihelion of 3.1803 AU, with an eccentricity of .08887. The inclination to the ecliptic is 10.884 degrees. Sylvia's orbit takes 2381.89 days, or 6.52 years. It rotates in 5.184 hours.

The surface gravity of Sylvia is 0.007G, and the escape velocity is 586 feet per second.

The density is a very low 1.2, suggesting that this asteroid may be a rubble pile. The albedo is a low .0435.

Sylvia is classified as a type X asteroid, with a surface mix of carbon and troilite (FeS).

* On February 18, 2001, Margot and Brown discovered a moon of Sylvia, given the name *Remus*. Remus is 438 miles from Sylvia, orbiting it in 1.37 days in an orbit inclined less than one degree to Sylvia's equator, with an eccentricity of .027.

* On August 9, 2004, Franck Marchis (b. 1973, asteroid 6639 Marchis) and colleagues discovered a second moon, given the obvious name *Romulus*. It is 841 miles from Sylvia, taking 3.65 days to orbit, with an inclination under one degree and an eccentricity of .006. This was the first discovery of an asteroid having more than one moon.

90 Antiope was discovered October 1, 1866, by K. Robert Luther (1822-1900; asteroid 1303 Luthera). It was one of 24 asteroids he discovered. On August 10, 2000, it was discovered that Antiope is actually two asteroids separated by 106 miles. The larger is 54.6 miles diameter, with a crater about 20 miles across. The smaller asteroid is 52.2 miles across.

* This pair has an aphelion of 3.6676 AU, and a perihelion of 2.6998 AU. The eccentricity of their barycenter around the Sun is .1593. The period is 2059.34 days, or 5.63 years. The two objects are in synchronous rotation of 16.5 hours.

The surface gravity is 0.003G, and escape velocity is 115 feet per second.

The albedo of this pair is .0603, and density of 1.25. They are class C.

Antiope is/are a member of the Themis family.

100 Hekate was discovered by James C. Watson (1838-1890; asteroid 729 Watson), the first Canadian born astronomer to discover an asteroid. The name is taken from the goddess of witches in ancient Greece. Hekate was the fourth of 22 asteroids Watson discovered. It has a diameter of 55 miles.

Hekate has an aphelion of 3.6068 AU, and a perihelion of 2.5685 AU. The eccentricity is .1681, with an inclination of 6.4301 degrees to the ecliptic. The period around the Sun is 1981.74 days, or 5.43 years. It rotates in 27.066 hours.

The surface gravity is 0.003G, and the escape velocity is 180 feet per second.

The albedo of Hekate is .1922, and density is 2.7. Hekate is class S.

108 Hecuba was discovered by Luther on April 2, 1869. It is named for the wife of King Priam during the Trojan War. Its diameter is 41 miles.

Hecuba has an aphelion of 3.4138 AU, a perihelion of 3.0587 AU, and an eccentricity of .0459. The orbital inclination is 4.223 degrees. The orbital period is 2126.49 days, or 5.82 years. It rotates in 17.86 hours.

The escape velocity is about 135 feet per second.

Hecuba is class S, with a fairly high albedo of .243.

Hecuba is the namesake of a small family of S class asteroids whose orbits partially overlap the much larger C class Hygiea family.

110 Lydia was discovered April 19, 1870, by Alphonse Borrelly (1842-1926; asteroid 1539 Borrelly), the second of his 18 asteroid discoveries. The name commemorates the land of the Phrygians in the ancient world. Lydia is 53.4 miles in diameter.

This asteroid has an aphelion of 2.9508 AU, and a perihelion of 2.5917 AU. The eccentricity is .0788 with an inclination of 5.9705 degrees. The orbital period is 1652.30 days, or 4.52 years. Lydia rotates in 10.927 hours.

The surface gravity is 0.0028G and escape velocity is 165 feet per second.

Lydia's albedo is .1808. The spectral class is M, suggesting a major amount of nickel and iron content.

120 Lachesis was discovered April 10, 1872, by Borrelly, his fourth asteroid. The diameter of 104.6 miles makes it his largest. Lachesis was one of the Greek fates handling destiny.

Lachesis has an aphelion of 3.2940 AU and a perihelion of 2.9424 AU. The eccentricity is .0564, and the inclination of the orbit is 6.949 degrees. The orbital period is 2011.20 days, or 5.51 years. Lachesis rotates in 46.551 hours.

The surface gravity is 0.0052G and the escape velocity is 305 feet per second.

The albedo of Lachesis is .0463, with a spectral class of C, as might be expected with such a low albedo.

121 Hermione was discovered by Watson on May 12, 1872. The shape is that of two spheres, 50 and 37 miles in diameter, mashed together. Hermione was the daughter of King Menelaus and Helen.

Hermione has an aphelion of 3.9142 AU and a perihelion of 2.9857 AU. The eccentricity is .13456, with an inclination of 7.5975 degrees. The orbital period is 2340.57 days, or 6.41 years. It rotates in 5.55 hours.

The surface gravity is 0.0027G. Escape velocity is 240 feet per second.

Hermione's albedo is .0482, and density is 1.9, which leads to its being class C.

* A moon eight miles in diameter was discovered by William Merline and colleagues in 2002. It has not received a formal name beyond *S/2002 (121) 1*, but the Merline group has proposed the name LaVictoire (the ship Lafayette rode on his first time in America was the Hermione; the ship for his second visit was the La Victoire.) It orbits 476 miles from Hermione in a period of 2.582 days. The escape velocity is 19 feet per second. It is 5.7 magnitudes fainter than Hermione suggesting, given the difference in sizes of the two bodies, that it has an albedo significantly brighter than that of its primary.

130 Elektra was discovered by Christian H. F. Peters (1806-1880) on February 17, 1873, number 18 of the 48 asteroids he discovered. (Peters is one of the few astronomers with large numbers of asteroids to his credit who does not have one named for him, perhaps because of a controversial legal battle he initiated late in his career.) The diameter is 113 miles. Elektra was the daughter of Agamemnon and Clytemnestra, and it is said mourning became her.

Elektra has an aphelion of 3.7765 AU, a perihelion of 2.4722 AU, and an eccentricity of .20872. The orbital inclination is 22.064 degrees. It orbits the Sun in 2017.14 days, or 5.52 years. Elektra rotates in 5.225 hours.

The surface gravity is 0.007G, with an escape velocity of 535 feet per second.

Elektra's albedo is .0755, with a spectral class of G and a density of 1.3.

* A moon with a diameter of 2.5 miles was discovered in an orbit 725 miles from Elektra on August 15, 2003. It is unnamed except for a catalog number, *S/2003 (130) 1.*

140 Siwa was discovered October 13, 1874, by Johann Palisa (1848-1925; asteroid 914 Palisana), one of 122 asteroids he discovered. Siwa was the Slavic goddess of fertility. Its diameter is 68 miles.

Siwa has an aphelion of 3.3243 AU, a perihelion of 2.1410 AU, and an eccentricity of .21651. The orbit is inclined 3.1865 degrees. It orbits the Sun in 1649.93 days, or 4.52 years. Siwa rotates in 34.407 hours.

The surface gravity is 0.0028G, with an escape velocity of 416 feet per second.

Siwa has an albedo of .0676, and is spectral class P.

145 Adeona was discovered June 3, 1875, by Peters. Adeona was the goddess of homecoming. The diameter is 94 miles.

Adeona's aphelion is 3.0631 AU, the perihelion is 2.2848 AU, and the eccentricity .1455. The inclination of the orbit to the ecliptic is 12.64 degrees. The orbital period is 1597.10 days, or 4.37 years. It rotates in 15.07 hours.

This class C asteroid has an albedo of .0433. The density is 1.18, with a surface gravity of 0.0048G and an escape velocity of 260 feet per second.

150 Nuwa was discovered by Watson on October 18, 1875. It was named for the Chinese goddess of creation. The diameter is 93.5 miles.

Nuwa has an aphelion of 3.3617 AU and a perihelion of 2.6051 AU, with an eccentricity of .1268. The inclination is 2.1932 degrees. It takes 1882.20 days, or 5.15 years to orbit the Sun. Nuwa rotates in 8.135 hours.

The albedo is .0395, low even for a Class C asteroid. The density is also quite low, 0.98, suggesting an object largely of rubble.

Nuwa is a member of the Hecuba family of asteroids.

153 Hilda was discovered November 2, 1875, by Palisa, and named for the daughter of his colleague, Theodor von Oppolzer. The diameter is 106 miles.

Hilda has an aphelion of 4.5228 AU, a perihelion of 3.4222 AU, and an eccentricity of .1385. The inclination is 7.8278 degrees. It takes 2891.98 days, or 7.92 years to orbit the Sun. It rotates in 5.96 hours.

The escape velocity is 19 feet per second. The albedo is .0618, fairly typical for a class C asteroid.

This is the namesake of the Hilda family of asteroids, typified by having a semimajor axis between 3.7 and 4.2 AU, an eccentricity less than .3, an inclination less than 20 degrees, and a 2:3 resonance with Jupiter.

156 Xanthippe was discovered November 22, 1875, by Palisa. It is named for the wife of Socrates (approximately 429 to 350 BC). It is 75 miles in diameter.

Xanthippe has an aphelion of 3.3462 AU, a perihelion of 2.1100 AU, and eccentricity of .2266. Orbital inclination is 9.776 degrees. The orbital period is 1645.87 days, or 4.51 years. It rotates in 22.37 hours.

This class C asteroid has an albedo of .0422. The surface gravity is 0.004G. The escape velocity is 200 feet per second.

158 Koronis was discovered January 4, 1876, by Victor Knorre (1840-1919; asteroid 14339 Knorre, shared with his father and grandfather). The diameter is 22 miles. It is one of four asteroids discovered by Victor Knorre. The asteroid's name commemorates the mother of Aesclepius.

The aphelion of Koronis is 3.0206 AU, the perihelion is 2.7164, and the eccentricity is .05303. The inclination is 1.0015 degrees. The orbital period is 1774.50 days, or 4.86 years. It rotates in 14.22 hours.

Koronis has an albedo of .2766, and is class S.

Koronis is the namesake of a family of asteroids characterized by semimajor axes between 2.85 and 2.9 AU, eccentricities of .035 and .09, and inclinations of 1 to 3.5 degrees. Over 300 are known.

160 Una was discovered by Peters on February 20, 1876. Una was a character in Spencer's *Faerie Queen.* The diameter is about 50 miles.

Una has an aphelion of 2.9047 AU, a perihelion of 2.5514 AU, and an eccentricity of .06476. The inclination is 3.8238 degrees. Una takes 1645.78 days to orbit the Sun, or 4.51 years.

The surface gravity is 0.003G, with an escape velocity of 144 feet per second.

Una is class C, with an albedo of .0645, and a density of 2.0.

170 Maria was discovered January 10, 1877, by Henri Joseph Perrotin. It was named for the sister of Antonio Abetti, who calculated its orbit. The diameter is 27 miles.

Maria has an aphelion of 2.7137 AU and a perihelion of 2.3943 AU. The eccentricity is .0625 and the inclination 14.384 degrees. The orbital period of Maria is 1490.87 days, or 4.08 years. It rotates in 13.12 hours.

This asteroid is class S, with an albedo of .1579.

Maria is the namesake of a grouping of asteroids with semimajor axes between 2.5 and 2.706 AU and inclinations of 12 to 17 degrees. Nearly one hundred members of this family are known.

180 Garumna was discovered January 29, 1878, by Perrotin and given the Latin name of the French river Garonne.

The aphelion is 3.1717 AU, the perihelion 2.2731 AU, and the eccentricity is .16503. The orbital inclination is 0.86635 degree. The orbital period is 1640.72 days, or 4.49 years. The rotation is 23.866 hours.

Garumna is class S.

190 Ismene was discovered by Peters on September 22, 1878. It is named for Antigone's sister (you forgot she had a sister?). Ismene has a diameter just under 100 miles.

Ismene has an aphelion of 4.6673 AU, a perihelion of 3.3345 AU, and an eccentricity of .16656. The inclination of the orbit to the ecliptic is 6.1552 degrees. The period is 2923.03 days, or 8.00 years. Ismene rotates in 6.155 hours.

This asteroid is class P and a member of the Hilda family. It is suspected of having some water ice.

200 Dynamene was discovered by Peters on July 27, 1879. The diameter is just under 80 miles. Dynamene was one of the fifty Nereids listed in ancient sources, which confusingly provides nearly a hundred names for the fifty. Nereids were minor deities in charge of rivers, springs, fountains, and other forms of flowing water. Their father was Nereus (see asteroid 4660).

Dynamene has an aphelion of 3.1034 AU, a perihelion of 2.3694 AU, and an eccentricity of .1341. The orbital inclination is 6.8966 degrees. It has an orbital period of 1653.38 days, or 4.53 years. It rotates in 37.4 hours.

216 Kleopatra was discovered April 10, 1880, by Palisa. It is 135 by 58 by 50 miles.

Kleopatra's aphelion is 3.4906 AU, the perihelion is 2.1000 AU, and eccentricity .2487. The orbital inclination is 13.10 degrees. The orbital period is 1707.05 days, or 4.67 years. It rotates in 5.385 hours.

The albedo is .1164, with spectral class M. The density is 4.27.

* Marchis found a three-mile-diameter moon named *Alexhelios*.

* Marchis found a 1.9 mile-diameter moon given the name *Clariselene*. Both names are based on the names of two of Cleopatra's children.

221 Eos was discovered January 18, 1882, by Palisa. The name is the ancient Greek for dawn. Its diameter is 65 miles.

Eos has an aphelion of 3.3243 AU, a perihelion of 2.6938 AU, and an eccentricity of .1048. The inclination of the orbit is 10.881 degrees. The orbital period is 1906.52 days, or 5.22 years. It rotates in 10.436 hours.

This asteroid has a K spectrum and an albedo of .140.

The members of this family have a semimajor axis between 2.99 and 3.83 AU. Eccentricities are .01 to .13. Orbits are inclined 8 to 12 degrees. Nearly 500 have been identified.

225 Henrietta was discovered April 19, 1882, by Palisa and named for the wife of one of his colleagues. The diameter is 75 miles.

Henrietta has an aphelion of 4.2886 AU, a perihelion of 2.4960 AU, and an eccentricity of .2642. The inclination of the orbit to the ecliptic is 20.868 degrees. The orbital period is 2282.12 days, or 6.25 years. It rotates in 7.36 hours.

The albedo is .0396. It is class T.

Henrietta is a member of the 65 Cybele family.

237 Coelestina was discovered June 27, 1884, by Palisa. He named it for the wife of Theodor von Oppolzer (1492 Oppolzer). The diameter is 25.5 miles.

The aphelion is 2.9657 AU, the perihelion is 2.5604, and the eccentricity .07335. The inclination of the orbit is 9.7481 degrees. The orbital period is 1677.58 days, or 4.59 years. It rotates in 29.215 hours.

This S class asteroid has an albedo of .2108.

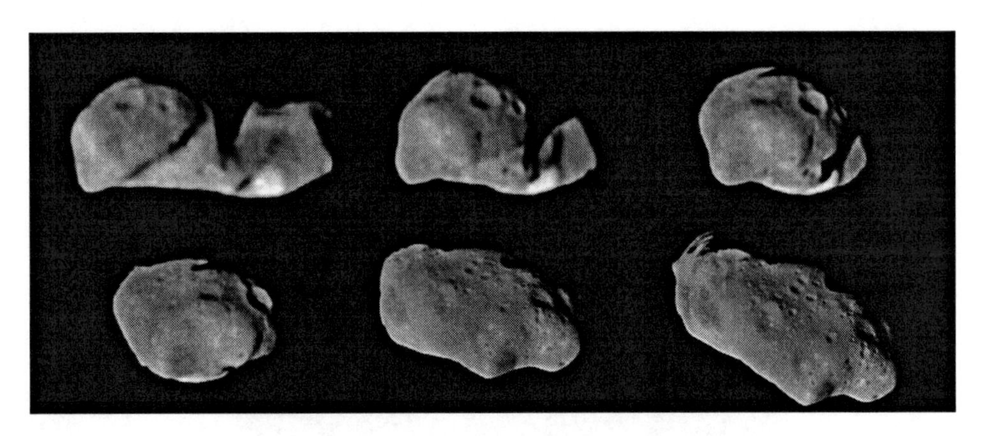

Asteroid Ida August 28, 1993, from 2200 miles.

243 Ida was discovered on September 29, 1884, by Palisa, his 45th asteroid. It is 36 by 14 miles. The name refers to an allegedly enchanted mountain in Greece.

Ida has an aphelion of 2.9814 AU, a perihelion of 2.7450 AU, and an eccentricity of .04163. The orbital period is 1768.69 days, or 4.84 years. It rotates in 4.634 hours.

The albedo is .2383. Ida is Class S. The density is 2.6.

The surface gravity is 0.0004G. Ida is a member of the Koronis family (158 Koronis).

* During the August 28, 1993, flyby of Ida by the Galileo probe on its way to Jupiter, a moon of Ida was discovered by Ann Horch. It was given the name Dactyl, the name of a race of sprites or fairies that inhabited Mount Ida in Greek myth. The diameter is 1.0 by .87 by .75 miles. Its orbital period is about 20 hours, with an orbit tilted eight degrees to Ida's equator. The surface appears to be saturated with craters.

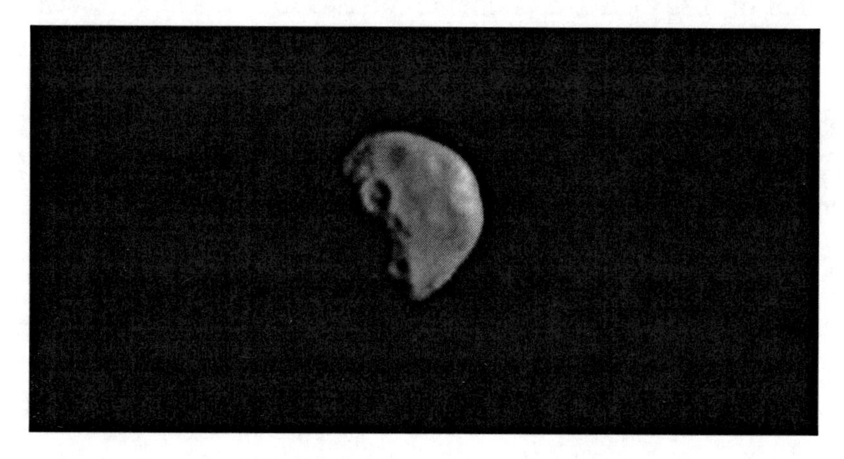

Asteroidal moon Dactyl, August 28, 1993 from 2400 miles. The large crater is about 1000 feet across.

250 Bettina was discovered September 3, 1885, by Palisa, who was paid fifty pounds by Baron Rothschild to name it for the Baron's wife. The diameter is 50 miles.

Bettina has an aphelion of 3.5602 AU, a perihelion of 2.7348 AU, and an eccentricity of .1311. The inclination is 12.81 degrees. It takes 2039.63 days, or 5.58 years, to orbit the Sun. The rotational period is 5.05 hours.

The albedo is .258. Bettina is class M.

Asteroid Matilda June 1997. The large crater is about 6 miles deep.

253 Mathilde was discovered November 12, 1885 by Palisa. The diameter is 32.5 miles. It is named for the wife of the then Vice Director of Paris Observatory, Moritz Loewy.

Mathilde has an aphelion of 3.35095 AU, a perihelion of 1.9422 AU, and an eccentricity of .26614. The inclination of the orbit to the ecliptic is 6.7413 degrees. The orbital period is 1572.64 days, or 4.31 years. It rotates in the very slow time of 417.7 hours (17.4 days).

The albedo is .0436. Mathilde is class C. Density is 1.3.

Surface gravity is 0.0097G, with an escape velocity of 76 feet per second.

On June 27, 1997, the spacecraft NEAR Shoemaker photographed Mathilde from as close as 753 miles, revealing many deep craters, with no hint of layering. Craters on Mathilde are named for coal fields on Earth.

260 Huberta was discovered October 3, 1886, by Palisa. It is named for Saint Hubertus (656-727). The diameter is 58.8 miles.

The aphelion is 3.8470 AU, the perihelion is 3.0474 AU, and the eccentricity .1160. The orbital inclination is 6.4141 degrees. The orbital period is 2337.77 days, or 6.40 years. Huberta has a 4:7 orbital resonance with Jupiter. It rotates in 8.29 hours.

This class C asteroid has an albedo of .0509. It is in the Cybele group.

275 Sapientia was discovered by Palisa on April 15, 1888. The name is Latin for wisdom. The diameter is 64 miles.

Sapientia has an aphelion of 3.2225 AU, a perihelion of 2.3027 AU, and an eccentricity of .16269. The inclination is 4.7633 degrees. The orbital period is 1685.33 days, or 4.61 years. The rotational period is 14.766 hours.

The albedo is .036, and it is a class C object. It has an escape velocity of 19 feet per second.

280 Philia was discovered October 29, 1888, by Palisa. The name is the Ancient Greek for friendship. The diameter is 28.4 miles.

The aphelion is 3.2586 AU, the perihelion 2.6297 AU, and the eccentricity .1068. The inclination of the orbit is 7.443 degrees. The orbital period is 1845.18 days, or 5.05 years. It rotates in 70.26 hours.

The albedo is .0444.

290 Bruna was discovered March 20, 1890, by Palisa. It is named for the Czech city of Brno, as is asteroid 2889. The diameter is about 13 miles.

Bruna's aphelion is 2.9391 AU, the perihelion 1.7366, and the eccentricity is .2572. The orbital period is 1305.63 days, or 3.57 years. It rotates in 13.807 hours.

300 Geraldina was discovered October 3, 1890, by Auguste Charlois (1864-1910, asteroid 1510 Charlois), the fourteenth of 99 asteroids he discovered before being murdered by a former brother-in-law. The diameter is 50 miles. The name may have commemorated a mistress or unacknowledged daughter.

Geraldina has an aphelion of 3.3871 AU, a perihelion of 3.0247 AU, and an eccentricity of .05653. The inclination is a very low 0.7324 degree. The period to revolve around the Sun is 2096.65 days, or 5.74 years. It rotates in 6.84 hours.

The albedo is .0397. Geraldina is class C.

323 Brucia was discovered December 22, 1891 by Maximilian Franz Josef Cornelius Wolf (usually just Max!; 827 Wolfiana). It is named for Catherine W. Bruce (1816-1900). This was the first asteroid discovered by photography. It is about 23 miles in diameter.

The aphelion is 3.0995 AU, the perihelion 1.6657 AU (just inside Mars's aphelion), and eccentricity .3009. The orbital inclination is 24.23 degrees. The orbital period is 1343.33 days, or 3.68 years. It rotates in 9.46 hours.

This class S asteroid has an albedo of .1765. The apparent magnitude ranges from 11.2 to 15.8. The surface gravity is 0.008G, with an escape velocity of 65 feet per second.

327 Columbia was discovered March 22, 1892, by Charlois, and is named for Christopher Columbus. The diameter is about 16.2 miles.

The aphelion of Columbia is 2.9523 AU, the perihelion is 2.6002 AU, and the eccentricity .0634. The inclination is 7.1485 degrees. The orbital period is 1689.64 days, or 4.68 years. It rotates in 5.93 hours.

The albedo is .2360.

400 Ducrosa was discovered on March 15, 1895, by Charlois. The diameter is about 31 miles. It was named for Joseph Ducros, a technical aide at the Nice Observatory where it was discovered.

Ducrosa's aphelion is 3.4899 AU, the perihelion is 2.7660 AU, and the eccentricity .1157. The inclination of the orbit is 10.53 degrees. The orbital period is 2020.62 days, or 5.53 years. It rotates in 6.87 hours.

The albedo of this S class asteroid is .1423.

Eros from 32 miles away, showing two overlapping craters.

433 Eros was discovered independently but simultaneously by Charlois and by Carl Gustav Witt (1846-1966, asteroid 2732 Witt) on August 13, 1898. It is 24.7 by 7 by 7 miles. It was the first asteroid found whose orbit was not wholly contained between Mars and Jupiter and was the first asteroid on which a spacecraft landed. NEAR-Shoemaker first went into orbit around Eros on February 14, 2000, mapping the surface. It landed on Eros on February 12, 2001, and conducted studies of the surface materials until finally ceasing transmissions on February 28, 2001.

Eros has an aphelion of 1.78245 AU, and a perihelion of 1.3322 AU, with an eccentricity of .22267. The inclination is 10.8287 degrees. The orbital period is 642.93 days, or 1.76 years. It rotates in 5.27 hours.

The albedo is .25, with a density of 2.67. It is class C, and in the Amor group.

With a wide range of distances from Earth, the apparent magnitude ranges from 7.0 to 15. The surface gravity is a mere 0.0006G, with an escape velocity of 40 feet per second.

Charlois Regio is the largest crater on Eros (for some reason NASA is reluctant to use the obvious adjectival form of the name, Erotic, instead using Erosian). It appears to be about one-billion years old and responsible for debris which has covered many smaller craters near it.

434 Hungaria was discovered September 11, 1898, by Max Wolf.

The aphelion is 2.0876 AU, the perihelion 1.8016 AU, and the eccentricity is .0735. The inclination of the orbit is 22.51 degrees. The orbital period is 990.45 days, or 2.71 years. It rotates in 26.51 hours.

This class E asteroid has an albedo of .428.

Hungaria is the namesake of the Hungaria family of asteroids, which are noted for semimajor axes between 1.78 and 2.00 AU, an eccentricity less than .18, an inclination between 16 and 34 degrees, a high albedo with enstatite, an important mineral, and orbital resonances of 9:2 with Jupiter and close to 3:2 with Mars.

452 Hamiltonia is the last asteroid discovered in the 1800s, December 6, 1899, by James E. Keeler (1857-1900; the first of his two discoveries; 2261 Keeler), who named it for Mount Hamilton, site of Lick Observatory where he made the discovery. It was promptly lost until 1981.

Hamiltonia has an aphelion of 2.8856 AU, a perihelion of 2.8092 AU, and an eccentricity of .0134. The inclination is 3.2249 degrees. The orbital period is 1754.97 days, or 4.80 years. It rotates in 2.88 hours, so it must be small or would be ripped apart by centrifugal force.

453 Tea was another Charlois discovery, on February 22, 1900, the first asteroid discovered in the 1900s. It has a 13-mile diameter. No one knows why he chose this name—perhaps a favored drink with a lady friend?

Tea has an aphelion of 2.4807 AU, a perihelion of 1.9452 AU, and an eccentricity of .1089. Inclination of the orbit to the ecliptic is 5.5507 degrees. It rotates on its axis in 6.81 hours.

The albedo is .189. Tea is a class S asteroid member of the Flora family (8 Flora).

480 Hansa was discovered May 21, 1901 by Max Wolf. It is named for the old Hanseatic League of German city states. The diameter is 35 miles.

Hansa's aphelion is 2.7637 AU, the perihelion is 2.5262 AU, and eccentricity .0449. The orbital inclination is 21.292 degrees. The orbital period is 1571.17 days, or 4.30 years. It rotates in 16.19 hours.

This is a class S asteroid with an albedo of .2485.

The Hansa family of asteroids is noted for a high inclination around 22 degrees and semimajor axis around 2.6 AU. Eccentricities are about .06. Fewer than a hundred have been identified.

499 Venusia was discovered December 24, 1902, by Max Wolf (1863-1932), just one of 248 asteroids of all types and classes that he discovered. The diameter is 50.6 miles.

The aphelion is 4.8925 AU, the perihelion is 3.1602 AU, and the eccentricity .2151. The inclination to the ecliptic is 2.0917 degrees. The orbital period is 2950.98 days, or 8.08 years. It rotates in 13.48 hours.

The albedo is .0468, and it is Class T.

500 Selinur was discovered January 16, 1903 by Wolf. The diameter is 26.8 miles. It is named for a character in a novel by Theodor Vischer.

Selinur's aphelion is 2.9939 AU, the perihelion is 2.2299 AU, and the eccentricity .1463. The inclination is 9.77 degrees. The orbital period is 1541.83 days, or 4.22 years. Selinur rotates in 8.01 hours.

This S class asteroid has an albedo of .1804.

550 Senta was discovered November 16, 1904, by Max Wolf. It is named for a town in Serbia. The diameter is 23.5 miles.

The aphelion is 3.1619 AU, the perihelion is 2.0212 AU, and eccentricity .2201. The orbital inclination to the ecliptic is 10.12 degrees. The orbital period is 1523.84 days, or 4.17 years. It rotates in 20.56 hours.

With an albedo of .2215, this is a class S asteroid.

588 Achilles was the first Trojan asteroid to be discovered on February 22, 1906, by Max Wolf. The diameter is 94 miles, making it the fourth largest known Trojan asteroid in Jupiter's L4 position. Achilles was one of the leading Greek warriors in the siege of Troy.

While the semimajor axis must be the same as Jupiter's as one of that planet's Trojans, Achilles has an aphelion of 5.9690 AU, and a perihelion of 4.4282 AU, with the eccentricity of .14820. Inclination of the orbit is 10.3175 degrees. The orbital period is necessarily the same as Jupiter's, 4329.40 days, or 11.85 years. Rotation is 7.3 hours.

The albedo is .0328, and it is Class D, like most Jovian Trojans. The density is 2.0.

The surface gravity is 0.004G, with an escape velocity of 95 feet per second.

600 Musa was discovered June 14, 1906, by Metcalfe. It is named for a Turkish mountain. The diameter is 15.3 miles.

Musa's aphelion is 2.8058 AU, the perihelion is 2.5138 AU, and eccentricity .0549. The inclination is 10.20 degrees. The orbital period is 1584.42 days, or 4.34 years. It rotates in 5.89 hours.

This class S asteroid has an albedo of .2415.

617 Patroclus was discovered October 17, 1906, by August Kopff (1882-1960; asteroid 1631 Kopff). The diameter is 87.6 miles.

Patroclus has an aphelion of 5.9460 AU, a perihelion of 4.4876 AU, and eccentricity .1398. The orbital inclination is 22.05 degrees. The orbital period is 4352.20 days, or 11.92 years. It rotates in the slow time of 102.8 hours.

Patroclus is class P, like most Jovian Trojans, with an albedo of .0471.

624 Hektor was discovered February 10, 1907, by August Kopff, number 22 of 68 asteroids he discovered. The diameter is 140 miles, making it the largest of the L4 "Greek" Trojans, and named for the only Trojan infiltrator into the Greek region. It is oblong, 230 by 125 by 120 miles.

Hektor has an aphelion of 5.3713 AU, a perihelion of 5.1205 AU, and an eccentricity of .02391. The inclination is 18.1748 degrees. It rotates in 6.924 hours.

It has an apparent magnitude ranging from 13.8 to 15.3. It is class D.

The density is 1.62. The surface gravity is 0.008G, with an escape velocity of 490 feet per second.

* A moon, *S/2006 (624) 1*, was discovered in orbit 620 miles away. It has a diameter of nine miles. This was the first moon discovered among Trojan asteroids

668 Dora was discovered July 27, 1908, by Kopff. The diameter is 16 miles.

Dora's aphelion is 3.4996 AU, the perihelion is 2.1427 AU, and eccentricity .2337. The orbital inclination is 6.841 degrees. The orbital period is 1707.83 days, or 4.68 years. It rotates in 22.914 hours.

This is a class C asteroid with an albedo of .0467.

700 Auravictrix was discovered June 5, 1910, by Joseph Helffrich (1890-1971, number 4 of 13 asteroids he discovered). The diameter is 9.5 miles.

The aphelion is 2.4604 AU, the perihelion is 1.9987 AU, and eccentricity .1035. The inclination of the orbit to the ecliptic is 6.7095 degrees. The orbital period is 1215.96 days, or 3.33 years. It rotates in 6.075 hours.

The albedo is .2455. It is a member of the Flora family.

729 Watsonia was discovered February 9, 1912, by Joel H. Metcalfe (1866-1925; the 28th of 41 asteroids he found; asteroid 792 Metcalfia). The diameter is about 30 miles.

The aphelion is 3.0243 AU, the perihelion is 2.4941 AU, and eccentricity .0961. The inclination of the orbit is 18.04 degrees. The orbital period is 1674.07 days, or 4.58 years.

This class B asteroid has an albedo of .1381, and rotates in 25.23 hours.

792 Metcalfia was discovered by Joel Metcalfe on March 20, 1907. He did not name it; that was done later to honor him. The diameter is about 38 miles.

The aphelion is 2.9630 AU, the perihelion is 2.2819 AU, and the eccentricity .1299. The inclination is 8.6169 degrees. The orbital period is 1551.12 days, or 4.25 years. It rotates in 9.17 hours.

This is a class B asteroid with an albedo of .0608.

800 Kressmannia was discovered March 20, 1915, by Max Wolf, who was being exceptionally practical by naming it for Edward Kressmann, a German wine merchant.

The aphelion is 2.6357 AU, the perihelion is 1.7495 AU, and eccentricity .2021. The orbital inclination is 4.265 degrees. The period is 1185.91 days, or 3.25 years. It rotates in 4.464 hours.

This class S asteroid is a member of the Flora group.

818 Kapteynia was discovered February 21, 1916, by Wolf. It is named for Jacobus Kapteyn (1851-1922), who has a star and a lunar crater also named for him. Many of his students and their students also have asteroids named for them. This book refers to such relationships as a *Collective*. The diameter is 31 miles.

Kapteynia has an aphelion of 3.4705 AU, a perihelion of 2.8657 AU, and an eccentricity of .09545. The orbit is inclined 15.6668 degrees to the ecliptic. The orbital period is 2059.67 days, or 5.64 years. It rotates in 16.35 hours.

The albedo is .1655.

@ This asteroid is the namesake of the Jacobus Collective of asteroids.

827 Wolfiana was discovered August 29, 1916, by Palisa. It is named for Maximilian Wolf.

Wolfiana has an aphelion of 2.6314 AU, a perihelion of 1.9187 AU, and an eccentricity of .1566. The orbit is inclined 3.421 degrees. The orbital period is 1253.40 days, or 3.43 years. It rotates in 4.07 hours.

Wolfiana is class S. It is a member of the Flora family.

832 Karin was discovered September 20, 1916 by Wolf. It is named for a Queen of Sweden, 1550-1612. The diameter is 21 miles.

Karin's aphelion is 3.0884 AU, the perihelion is 2.6405 AU, and the eccentricity .0782. The inclination is 1.0056 degrees. The orbital period is 1770.76 days, or 4.85 years. It rotates in 18.35 hours. It is S class.

The Karin family of asteroids, all of which are considerably smaller than Karin, has been traced back to a collision which disrupted the impacting bodies 5.8 +/- 0.02 million years ago. This makes the Karin family the youngest known asteroid family. It was first identified in 2002. Karin was also one of the first asteroids to demonstrate the Yarkovsky Effect and is still the largest to show it.

837 Schwarzschilda was discovered September 23, 1916, by Max Wolf. It is named for astrophysicist Karl Schwarzschild (1873-1916).

The aphelion is 2.3925 AU, the perihelion is 2.2045 AU, and the eccentricity .0409. The orbital inclination to the ecliptic is 6.732 degrees. The orbital period is 1272.80 days, or 3.48 years. It may rotate in about 24 hours.

887 Alinda was discovered January 3, 1918, by Max Wolf. It has a diameter of about three miles.

Alinda has an aphelion of 3.8845 AU, a perihelion of 1.0713 AU, and an eccentricity of .56767. The period is 1424.71 days, or 3.90 years. The orbit is inclined 9.3592 degrees. It rotates in 73.97 hours. Alinda is in almost a 3:1 resonance with Jupiter and 1:4 with Earth.

This asteroid is class S, with an albedo of .31.

On January 20, 2025 Alinda will pass 7.63 million miles from Earth.

Alinda is the namesake of a small family of two dozen asteroids having a semimajor axis of around 2.5 AU and eccentricities of .4 to .65. They all have similar resonances to Earth and Jupiter, making their orbits chaotic on scales of tens of thousands of years.

944 Hidalgo was discovered October 31, 1920, by Walter Baade (1893-1960; asteroid 1501 Baade), the first of ten asteroids he found. It is named for Don Miguel Hidalgo y Costillo, who initiated the Mexican War of Independence with the Cry of Dolores. The diameter is 23.3 miles.

Hidalgo has an aphelion of 9.5337 AU, a perihelion of 1.9404 AU, and an eccentricity of .6618. The period is 5019.11 days, or 13.74 years. The inclination of the orbit is 42.54 degrees. The rotation is 10.06 hours.

Hidalgo is a class D asteroid with an albedo of .06. It falls into the family of centaurs, the first such discovered decades before this family was defined.

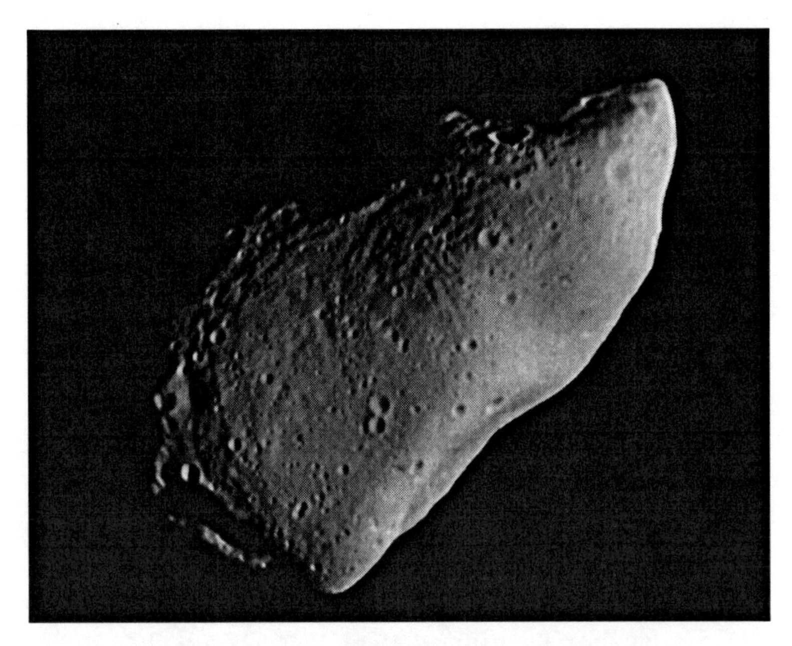

Asteroid Gaspra October 29, 1991 from 3300 miles.

951 Gaspra was discovered July 30, 1916, by Grigory Neujmin (1886-1946; asteroid 1129 Neujmin). It is one of 74 asteroids that he discovered. The diameter is 7.5 miles. The name honors a popular Black Sea resort in Ukraine.

The aphelion is 2.5932 AU, perihelion 1.8253 AU, and eccentricity is .17378. The period is 1199.40 days, or 3.28 years. Inclination is 4.103 degrees. Rotation is 7.042 hours.

Gaspra has an albedo of .22 and a density of 2.7. It is class S.

Surface gravity is 0.003G, with an escape velocity of 20 feet per second.

Gaspra was photographed from close up by the Galileo probe on its way to Jupiter on October 29, 1991.

989 Schwassmannia was discovered November 18, 1922, by Schwassmann himself (Arnold Schwassmann, 1870-1964) and later named by the IAU for him. He is better known for co-discovery of several comets, although he also found 22 asteroids. The diameter is eight miles.

Schwassmannia has an aphelion of 3.3248 AU, a perihelion of 1.9949 AU, and eccentricity of .2500. The orbital inclination is 14.70 degrees. The orbital period was 1584.47 days, or 4.34 years. It rotates in 4.58 hours.

998 Bodea was discovered August 6, 1923, by Karl W. Reinmuth (asteroid 1111 Reinmuth). It is named for Johann Bode, famous for Bode's Law. The diameter is 23.7 miles.

The aphelion is 3.7334 AU, the perihelion is 2.4662 AU, and eccentricity .2108. The inclination of the orbit is 15.49 degrees. The orbital period is 2017.98 days, or 5.52 years. It rotates in 8.574 hours.

The albedo is an extremely low .0211, suggesting a C class.

1000 Piazzia was discovered August 12, 1923, by Karl Reinmuth. It is named for Giuseppe Piazzi, who discovered the first asteroid, Ceres. The diameter is 29.7 miles.

Piazzia has an aphelion of 3.9867 AU, a perihelion of 2.3574 AU, and an eccentricity of .25683. The inclination is 20.57 degrees. The period is 2063.52 days, or 5.65 years. It rotates in 9.47 hours.

Surface gravity is 0.006G, with an escape velocity of 91 feet per second. The density is about 2.0, with an albedo of .1119. It is probably class S.

1001 Gaussia was discovered August 8, 1923, by S. Belyovski. It is named for Carl F. Gauss (1777-1855). The diameter is 46.4 miles.

The aphelion is 3.6191 AU, the perihelion is 2.8008 AU, and the eccentricity .1275. The orbital inclination is 9.297 degrees. Orbital period is 2100.63 days, or 5.75 years. It rotates in 9.17 hours.

The albedo is .0392, and is class P.

1002 Olbersia was discovered August 15, 1923, by Vladimir Albitzski (1891-1952), the first of ten asteroids he discovered. It is named for Heinrich Wilhelm Olbers, who discovered two of the earliest known asteroids and created Olbers's Paradox regarding the age and size of the universe. The diameter is 20 miles.

The aphelion is 3.2141 AU, the perihelion 2.3675 AU, and eccentricity .1517. The inclination of the orbit is 10.75 degrees. The orbital period is 1702.90 days, or 4.66 years. It rotates in 10.24 hours.

The albedo is .0621, and density is around 2. The surface gravity is 0.001G. The escape velocity is 60 feet per second.

1036 Ganymed was discovered October 23, 1924, by Baade, the sixth of his ten asteroids. It shares its name with the largest moon in the Solar System, Ganymede, remembering the cup bearer of the gods on Mount Olympus (Baade used the German spelling of the name). The diameter is 21.3 miles, making it the largest Near Earth Asteroid.

The aphelion is 4.0838 AU, the perihelion 1.2406 AU, and eccentricity .5430. The inclination is 26.70 degrees. The orbital period is 1586.59 days, or 4.34 years. It rotates in 10.31 hours.

It is S class, with a surface of orthopyroxenes and magnesium silicates. The albedo is .2926, and the density 2.9. Ganymed has an escape velocity of 56 feet per second, and a surface gravity of 0.0009G. It is a member of the Amor group.

1069 Planckia was discovered by Wolf on January 2, 1927. It is named for Max Planck (1858-1947), who also has a constant. The diameter is 27 miles.

Planckia's aphelion is 3.4704 AU, the perihelion is 2.7841 AU, and eccentricity .1097. The inclination of the orbit is 13.570 degrees. The orbital period is 2019.97 days, or 5.53 years. It rotates in 8.665 hours.

This class S asteroid has an albedo of .2158.

1089 Tama was discovered November 17, 1927, by Okuro Oikawa (1896-1970; 8 asteroids discovered; 2667 Oikawa). It is named for a river in Japan. The diameter is 8.1 miles.

Tama's aphelion is 2.4953 AU, the perihelion 1.9321 AU, and eccentricity .1272. The orbit's inclination is 3.7208 degrees. The orbital period is 1203.02 days, or 3.29 years.

The density is 2.52, with an albedo of .24. The surface gravity is 0.004G, with an escape velocity of 22 feet per second.

*_*S/2003* (*1089*) *1* was discovered December 24, 2003. It has a diameter of six miles and orbits Tama at a distance of 12.5 miles in 0.852 day, or 16.44 hours. Rotation and revolution are synchronous.

1111 Reinmuth was discovered February 11, 1927, by Reinmuth himself, leading to a change in the rules of asteroidal naming to prevent a recurrence (but see asteroids 1832 and 2325 for an example of a loophole).

The aphelion is 3.2831 AU, the perihelion 2.7050 AU, and eccentricity .0965. The inclination of the orbit is 3.885 degrees. The orbital period is 1892.31 days, or 5.18 years. It rotates in 4.02 hours.

This is a class T asteroid.

1129 Neujminia was discovered August 8, 1929, by Praskoniya Parchomenko. It is named for Grigory Neujmin. The diameter is 21.6 miles.

The aphelion is 3.2651 AU, the perihelion is 2.7820, and the eccentricity .07988. The orbital inclination is 8.6167 degrees. The period of revolution around the Sun is 1920.32 days, or 5.26 years. It rotates in 5.08 hours.

This class T asteroid has an albedo of .1216.

1134 Kepler was discovered September 25, 1929 ,by Max Wolf. It is named for Johannes Kepler, who discovered the laws of planetary motion.

Kepler's aphelion is 3.9318 AU, the perihelion is 1.4385, and the eccentricity is .4643. The orbital inclination is 6.3203 degrees. The period of the orbit is 1607.172 days, or 4.40 years. It crosses the orbit of Mars.

This is an S class asteroid.

1143 Odysseus was discovered by Reinmuth on January 28, 1930. It is named for the hero of Homer's Illiad and Odyssey. The diameter is 68 miles.

Odysseus has an aphelion of 5.7272 AU and a perihelion of 4.7710 AU with an eccentricity of .09108. The inclination is 3.138 degrees. The period is 4392.70 days or 12.03 years. It rotates is 10.11 hours.

The albedo is .0753 with an apparent magnitude of 7.93. Odysseus is one of Jupiter's Trojans, in the L4 position.

1202 Marina was discovered September 13, 1931, by Neujmin. The diameter is 34 miles.

Marina has an aphelion of 4.6555 AU, a perihelion of 3.3277 AU, and an eccentricity of .1663. The orbital inclination is 3.3357 degrees. The orbital period is 2912.81 days, or 7.97 years. It rotates in 9.45 hours.

The albedo is .0337.

1215 Boyer was discovered January 19, 1932 by Alfred Schmitt (asteroid 1617 Schmitt). It is named for Louis Boyer, who named an asteroid for Schmitt.

The aphelion is 2.9169 AU, the perihelion is 2.2402 AU, and the eccentricity .1312. The inclination of the orbit is 15.92 degrees. The orbital period is 1512.40 days, or 4.14 years.

It is class S.

1272 Gefion was discovered October 10, 1931, by Reinmuth. It is named for the Norse goddess of the plow.

Gefion's aphelion is 3.2087 AU, the perihelion is 2.3571 AU, and eccentricity .1530. The orbital inclination is 8.422 degrees. The orbital period is 1695.69 days, or 4.64 years.

Its spectrum is class B.

Asteroids in Gefion's family have a semimajor axis of 2.74 to 2.82 AU, an eccentricity between .12 and .148, and inclinations of 8.4 to 9.6 degrees. About a hundred have been identified.

1303 Luthera was discovered March 16, 1928 by Schwassmann. It is named for K. Robert Luther. The diameter is about 53 miles.

The aphelion is 3.5791 AU, the perihelion is 2.8876 AU, and the eccentricity is .10697. The inclination of the orbit is 19.51 degrees. The orbital period for revolving around the Sun is 1810.80 days, or 4.96 years. It rotates in 5.88 hours.

The albedo is .0608.

1313 Berna was discovered August 24, 1933 by Sylvain J. V. Arend (1902-1992, the eighth of 51 asteroids he discovered). It is named for the city of Berne, Switzerland. It is 6.5 miles in diameter.

The aphelion is 3.1095 AU, the perihelion 2.1005 AU, and the eccentricity is .2088. The inclination is 1.061 degrees. The orbital period is 1580.14 days, or 4.33 years. It rotates in 25.46 hours.

The density is 1.21, with an albedo of about .1. The surface gravity is 0.00065G. The escape velocity is 14 feet per second.

* A moon was discovered February 12, 2004 by Stefano Sposetti (b. 1958). It has a seven mile diameter, and at a distance of 22 miles orbits Berna in 1.06 days. It only has a catalog number of *S/2004* (*1313*) *1*.

1450 Raimonda was discovered February 20, 1938, by Yrjo Vaisala (1891-1971; discovered 128 asteroids and has two named for him: 1573 Yrjo and 2804 Vaisala). It is 9.5 miles in diameter. The name honors Jean-Jacques Raimond, Jr.

Raimonda has an aphelion of 3.0616 AU, a perihelion of 2.1619 AU, and an eccentricity of .1722. The orbit is inclined 4.866 degrees to the ecliptic. The orbital period is 1541.67 days, or 4.22 years. It rotates in 12.66 hours.

The albedo is .1387, suggesting S class. This asteroid is a member of the Jacobus Collective.

1492 Oppolzer was discovered March 23, 1938 by Y. Vaisala. It is named for Theodor von Oppolzer, who wrote what was the standard work on eclipses until Fred Espenak's (14120 Espenak) work in the early years of the twenty-first century. The diameter is 7.5 miles.

The aphelion is 2.4269 AU, the perihelion is 1.9182 AU, and the eccentricity is .1171. The inclination of the orbit is 6.056 degrees. The orbital period is 1169.66 days, or 3.20 years.

The albedo is .0890.

1501 Baade as discovered October 20, 1938, by Arthur Wachmann (1902-1990), the second of three asteroids he discovered. It is named for the prominent German astronomer Walter Baade (1892-1960). The diameter is about 7 miles.

The aphelion is 3.1584 AU, the perihelion 1.9288 AU, and the eccentricity is .2417. The orbital inclination is 7.322 degrees. The orbital period is 1481,74 days, or 4.06 years. It rotates in 15.13 hours.

The albedo is .2093.

1566 Icarus was discovered June 27, 1949, by Baade. Icarus was the youth who flew too near the Sun and fell to his death. It is the eighth of Baade's ten asteroids. The diameter is about 3000 feet.

Icarus has an aphelion of 1.9693 AU, a perihelion of 0.18657 AU (closer to the Sun than Mercury ever gets), and an eccentricity of .8269. The inclination is 22.83 degrees. The orbital period is 408.78 days, or 1.12 years. It rotates in 2.27 hours.

The albedo is .51, with a density of 2. The surface gravity is 0.00004G, and the escape velocity is 2.5 feet per second.

Icarus is a member of the Apollo, Alinda, and Aten groups by most definitions.

1573 Vaisala was discovered October 27, 1949, by Arend. It is named for Yrjo Vaisala (1891-1971, a Finnish astronomer who also has 2804 Yrjo named for him). The diameter is six miles.

Vaisala's aphelion is 3.2419 AU, the perihelion is 1.8211 AU, and eccentricity .2317. The inclination of the orbit to the ecliptic is 24.55 degrees. The orbital period is 1332.80 days, or 3.65 years. It rotates in 2.52 hours.

The albedo is .2226.

1578 Kirkwood was discovered January 10, 1951, by a group headed by B. Potter at Indiana University, where Kirkwood had taught. The diameter is 32 miles.

1578 Kirkwood has an aphelion of 4.8758 AU, a perihelion of 2.9960 AU, and an eccentricity of .23881. The inclination is 0.8107 degree. The orbital period is 2852.09 days, or 7.81 years. It rotates in 12.5 hours.

It is Class D, with an albedo of .0517.

1604 Tombaugh was discovered March 24, 1931, by Carl O. Lampland. It was named on the fiftieth anniversary of the discovery of Pluto for Pluto's discoverer, Clyde W. Tombaugh (1906-1997, found over 800 asteroids). The diameter is 20 miles.

Tombaugh's aphelion is 3.3257 AU, the perihelion is 2.7330 AU, and eccentricity .0978. The orbit is inclined 9.3865 degrees. The orbital period is 1925.89 days, or 5.27 years. It rotates in 7.047 hours.

The spectral class is B, with an albedo of .1038.

1617 Schmitt was discovered March 20, 1952, by Louis Boyer (1901-1999; asteroid 1215 Boyer, discovered 40 asteroids). It is named for Alfred Schmitt.

The aphelion is 3.6169 AU, the perihelion is 2.7819 AU, and the eccentricity .1305. The inclination of the orbit is 13.24 degrees. The orbital period is 2090.25 days, or 5.72 years. It rotates in 7.062 hours.

1620 Geographos was discovered September 14, 1951, by Albert Wilson (1918-2012), the first of five asteroids he discovered. It is named for the National Geographic Society. It is 3 by 1.3 by 1.2 miles.

The aphelion is 1.6635 AU, the perihelion is 0.8277 AU, with an eccentricity of .3355. The inclination of the orbit is 13.34 degrees. The orbital period is 507.77 days, or 1.39 years. It rotates in 5.22 hours.

This S class asteroid has a density of 2.0, and an albedo of .3258. The surface gravity is 0.00009G, with an escape velocity of six feet per second. It is in the Apollo group.

1622 Chacornac was discovered March 15, 1952, by Alfred Schmitt (1907-1973; 1617 Schmitt; the second of four he found). It is named for Jean Chacornac, a French astronomer who discovered six asteroids and a comet.

The aphelion is 2.6006 AU, the perihelion 1.8678 AU, and the eccentricity .1640. The inclination is 6.4590 degrees. The orbital period is 1219.78 days, or 3.34 years. It rotates in 12.216 hours.

1631 Kopff was discovered October 11, 1936, by Vaisala. It is named for August Kopff (1882-1960; found 68 asteroids). The diameter is about 6 miles.

Kopff's aphelion is 2.7110 AU, the perihelion is 1.7597 AU, and eccentricity is .2128. The orbital inclination is 7.489 degrees. The orbital period is 1220.74 days, or 3.34 years. It rotates in 6.683 hours.

The albedo is .2497.

1635 Bohrmann was discovered March 7, 1924, by Reinmuth and was named for Alfred Bohrmann.

The aphelion is 3.0178 AU, the perihelion is 2.6873 AU, and the eccentricity is .0579. The inclination of the orbit is 1.8158 degrees. The orbital period is 1759.74 days, or 3.76 years. It rotates in 5.864 hours.

This is another S class.

1666 VanGent was discovered July 22, 1930, by van Gent.

The aphelion is 2.5845 AU, the perihelion is 1.7854 AU, and the eccentricity .1829. The inclination is 2.687 degrees. The orbital period is 1179.64 days, or 3.23 years.

1685 Toro was discovered on July 17, 1948, by Carl A. Wirtanen (1910-1990; the second of eight asteroids he found). It is named for the wife of astronomer Samuel Herrick. The diameter is 2.0 miles.

The aphelion is 1.9632 AU, the perihelion is 0.7711, and eccentricity is .4360. The inclination of the orbit is 9.38 degrees. The period is 583.89 days, or 1.60 years. It rotates in 10.2 hours.

This class S Apollo asteroid is in a 5:8 resonance with Earth and a 5:13 resonance with Venus. It crosses the orbits of Mars and Earth, and approaches that of Venus. The albedo is .31.

1686 DeSitter was discovered September 28, 1935, by Hendrick van Gent (1900-1997; asteroid 1666 vanGent; the fifteenth of 39 he found). It is named for Willem De Sitter. The diameter is 19 miles.

This asteroid has an aphelion of 3.6792 AU, a perihelion of 2.6441 AU, and an eccentricity of .1367. The inclination is 0.6331 degree. The orbital period is 2053.38 days, or 5.62 years.

It is a member of the Jacobus Collective.

1691 Oort was discovered September 9, 1956, by Ingrid van Houten-Groeneveld and Karl Reinmuth. It was named for Jan Oort.

The aphelion is 3.705 AU, the perihelion is 3.6262 AU, and eccentricity is .1704. The inclination of the orbit is 1.0783 degrees. The orbital period around the Sun is 2057.21 days, or 5.63 years. It rotates in 10.27 hours.

This asteroid is believed to be class T or C. It is a member of the Jacobus Collective.

1700 Zvezdara was discovered August 27, 1940, by Pera Durkovic, one of two asteroids he discovered. It is named for the location of the observatory where it was discovered (*zvezda* is Russian for star). The diameter is 13 miles.

Aphelion of Zvezdara is 2.8915 AU, the perihelion is 1.8304 AU, and the eccentricity is .2247. The inclination is 4.512 degrees. The orbital period is 1325.05 days, or 3.63 years. It rotates in 9.114 hours.

The albedo is .0425 and it may be class T.

1745 Ferguson was discovered September 17, 1941, by J. Willis. It is named for James Ferguson.

Ferguson's aphelion is 3.001 AU, the perihelion is 2.6898 AU, and the eccentricity is .0546. Inclination of the orbit is 3.2580 degrees. The orbital period is 1752.94 days, or 4.80 years.

1762 Russell was discovered October 8, 1952, by B. Potter. It is named for Henry Norris Russell (1877-1957), co-developer of the HR Diagram of stellar types.

Russell's aphelion is 3.1024 AU, the perihelion is 2.6524, and eccentricity of .0782. The orbital inclination is 2.28 degrees. The orbital period is 1782.76 days, or 4.88 years.

1776 Kuiper was discovered September 24, 1960, by Cornelius J. van Houten, Ingrid van Houten-Groeneveld, and Tom Gehrels. The diameter is 22.3 miles. It is named for Gerard P. Kuiper, who also has a lunar crater and the Kuiper Belt named for him.

The aphelion is 3.1453 AU, the perihelion is 3.0632 AU, and the eccentricity .01322. The inclination is 9.49 degrees. The orbital period is 1997.72 days, or 5.47 years.

This asteroid has an albedo of .0544. It is a member of the Jacobus Collective.

1781 Van Biesbroeck was discovered October 17, 1906, by Kopff. It is named for George van Biesbroeck (1880-1974; found 16 asteroids). He also has a star named for him, otherwise GJ 752B.

The aphelion is 2.6517 AU, the perihelion is 2.1388 AU, and eccentricity is .1071. The orbital inclination to the ecliptic is 6.947 degrees. The orbital period around the Sun is 1354.00 days, or 3.71 years.

1795 Woltjer was discovered September 24, 1960, by Houten, Houten-Groeneveld and Gehrels. It is named for Jan Woltjer. The diameter is 16.8 miles.

The aphelion is 3.3176 AU, the perihelion is 2.2653 AU, and the eccentricity .1885. The inclination is 7.521 degrees. The orbital period is 1703.50 days, or 4.66 years.

The albedo is .0459. Geologically it is class B. It is a member of the Jacobus Collective.

1823 Gliese was discovered September 4, 1951, by Reinmuth. It is named for Wilhelm Gliese (1915-1993), who was instrumental in the study of stars near the Solar System.

Gliese's aphelion is 2.5268 AU, the perihelion is 1.9242 AU, and the eccentricity .1354. Orbital inclination to the ecliptic is 2.893 degrees. The orbital period is 1212.68 days, or 3.32 years.

1832 Mrkos was discovered August 11, 1969, by Lyudmila Chernykh (b. 1935; asteroid 2325 Chernykh). It honors the prominent Czech astronomer Antonin Mrkos (1918-1996). The diameter is 19.1 miles.

The aphelion is 3.5525 AU, the perihelion is 2.8741 AU, and the eccentricity .10556. The orbital inclination is 14.95 degrees. The orbital period is 2103.87 days, or 5.76 years. It rotates in 13.64 hours.

The albedo is .0742.

1850 Kohoutek was discovered March 23, 1942, by Reinmuth. It is named for Czech astronomer Lubos Kohoutek (b. 1935, discovered 75 asteroids).

Kohoutek's aphelion is 2.5350 AU, the perihelion is 1.9662 AU, and eccentricity .1264. The inclination of the orbit is 4.051 degrees. The orbital period is 1233.27 days, or 3.38 years.

1855 Korolyev was discovered October 8, 1969, by Lyudmila Chernykh. It honors Sergey Korolyev (1907-1966), who, until his untimely death from cancer, was head and inspiration of the Soviet Space Program under the title and pseudonym of Chief Designer.

The aphelion is 2.4374 AU, the perihelion is 2.0564 AU, and the eccentricity .08478. The inclination is 3.0785 degrees. The orbital period is 1230.22 days, or 3.37 years. It rotates in 4.66 hours.

1862 Apollo was discovered April 24, 1932, by Reinmuth. It is named for the Greek god of the Sun. While irregularly shaped, the mean diameter is 0.9 mile.

The aphelion is 2.2931 AU, the perihelion 0.6471 AU, and the eccentricity .5598. This orbit crosses the orbits of Mars, Earth, and Venus. The orbital inclination is 6.354 degrees. The orbital period is 651.051 days, or 1.78 years. It rotates in 3.065 hours.

A class Q asteroid, the albedo is .25. Density is 2.0.

* On November 4, 2005, the Arecibo radio telescope detected a moon, *S/2005* (*1862*) *1*. With a diameter of 260 feet, it orbits only 1.8 miles from Apollo.

1877 Marsden was discovered March 24, 1974, by van Houten, van Houten-Groeneveld, and Gehrels. It is named for Brian Marsden (1937-2010), who was involved in the discovery of 25 of Saturn's moons and was the long-time chair of the IAU's small-body naming commission. He co-discovered asteroid 37556 Svyaztie.

Marsden's aphelion is 4.7645 AU, the perihelion is 3.1191 AU, and the eccentricity is .2095. The inclination of the orbit is 17.556 degrees. The orbital period is 2855.82 days, or 7.82 years. It rotates in 14.4 hours.

1905 Ambartsumian was discovered May 14, 1972, by Tamara Smirnova (asteroid 5540). It is named for Viktor Ambartsumian (1908-1996), founder and head of Byurakan Observatory, who also served as IAU President.

Ambartsumian's aphelion is 2.5849 AU, the perihelion is 1.8612 AU, and eccentricity .1628. The orbital inclination is 2.616 degrees. The orbital period is 1210.68 days, or 3.31 years.

1911 Schubart was discovered October 25, 1973, the twentieth of 94 found by Paul Wild. It is named for Joachim Schubart (b. 1928). The diameter is 50 miles.

Schubart's aphelion is 4.6417 AU, the perihelion is 3.3185 AU, and the eccentricity .1662. The inclination is 1.65 degrees. The orbital period is 2900.27 days, or 7.94 years.

The albedo is a low .0249 for this class P asteroid.

1932 Jansky was discovered October 27, 1971, by Lubos Kohoutek. It is named for Karl Jansky (1905-1950), who discovered radio emissions from the center of our galaxy, essentially founding radio astronomy, which was ignored by many astronomers in the 1930s. As late as 1957 I was warned to avoid radio astronomy as a passing fad of no importance. HA!

Jansky's aphelion is 2.7476 AU, the perihelion is 1.9957 AU, and eccentricity .1585. The inclination of the orbit is 1.89 degrees. The orbital period is 1334.06 days, or 3.65 years.

This is a class B asteroid.

1951 Lick was discovered July 26, 1949 by Wirtanen. It is named for Lick Observatory, where he worked, and which itself is named for James Lick, who donated the mountain to the University of California and is buried beneath one of the telescopes. The asteroid's diameter is 3.6 miles.

Lick's aphelion is 1.4762 AU, the perihelion is 1.3905 AU, and eccentricity .0616. The inclination of the orbit to the ecliptic is 39.09 degrees. The orbital period is 598.91 days, or 1.63 years. The rotational period is 5.3 hours. The orbit crosses that of Mars.

The albedo is .0895. It is class A, with olivine, pyroxene, and metals.

1952 Chaos was discovered November 19, 1998 by the Deep Ecliptic Survey. The diameter is about 360 miles.

The aphelion of Chaos is 51.086 AU, the perihelion is 41.923 AU, and eccentricity .1093. The orbital inclination is 12.019 degrees. The orbital period is 114159.42 days, or 312.55 years. It rotates in 3.985 days.

The albedo is .05.

1958 Chandrasekhar was discovered September 24, 1970, by Carlos Cesco at the Yale-Columbia Southern Observatory. It is named for Subrahmanyan Chandrasekhar (1919-1995), the first astronomer to win a Nobel Prize (in physics) for explaining how the Sun and other Main Sequence stars make energy. The diameter is 21.1 miles.

The aphelion is 3.6247 AU, the perihelion 2.5829 AU, and the eccentricity .1678. The inclination of the orbit is 10.56 degrees. The orbital period is 1997.29 days, or 5.47 years.

The albedo is .0801.

1983 Bok was discovered on June 9, 1975, by Elizabeth Roemer (b. 1929; one of two asteroids she has discovered). It is named for Bart J. Bok (1906-1983), who announced when it was named that he planned to retire there, and his wife Priscilla.

The aphelion is 2.8805 AU, the perihelion 2.3639 AU, and the eccentricity is .0985. The orbital inclination is 9.407 degrees. It revolves around the Sun in 1550.95 days, or 4.25 years.

This asteroid is a member of the Jacobus Collective.

1991 Darwin was discovered May 6, 1967, by Carlos Cesco (discoverer of 20 asteroids, d. 1987) and Arnold Klemola. It is named for Charles Darwin (1809-1882).

Darwin's aphelion is 2.7158 AU, the perihelion 1.7812 AU, and eccentricity .2078. The inclination of the orbit to the ecliptic is 5.919 degrees. The orbital period is 1231.49 days, or 3.37 years.

1998 Titius was discovered February 24, 1958, by Alfred Bohrmann, the fifth of nine asteroids he discovered. It is named for Johann Titius, who was the first to come up with what was later called Bode's Law. So, not only is it named for the wrong person, it isn't even a valid law!

The aphelion is 2.5722 AU, the perihelion 2.2650 AU, and the eccentricity is .0635. The inclination of the orbit is 7.6291 degrees. The period of the orbit is 1373.87 days, or 3.76 years. It is in a 2:1 resonance with Mars, and rotates in 6.13 hours.

This asteroid is class B, with an albedo of .107.

2000 Herschel was discovered July 29, 1960, by Joachim Schubart (b. 1928; one of two asteroids he has discovered—asteroid 1911 Schubart—and who is a noted specialist in the study of asteroids with orbital resonances, particularly the Hilda group). It is named for William Herschel, who discovered Uranus and two of its moons.

The aphelion is 3.0882 AU, the perihelion is 1.6716 AU, and the eccentricity is .2976. The inclination is 22.798 degrees. The orbital period is 1341.05 days, or 3.67 years. It rotates in 130 hours.

The albedo is approximately .1 and is believed to be class S. Since its perihelion is barely inside the aphelion of Mars's orbit, it is technically a member of the Aten group.

2018 Schuster was discovered October 17, 1931, by Reinmuth. It is named for Hans-Emil Schuster (b. 1934; discovered 25 asteroids).

Schuster's aphelion is 2.6037 AU, the perihelion is 2.1833 AU, and the eccentricity is .1926. The inclination of the orbit to the ecliptic is 2.558 degrees. The orbital period is 1178.32 days, or 3.23 years.

2044 Wirt was discovered November 8, 1950, by Carl Wirtanen, for whom it is named (1910-1990; sixth of eight asteroids he discovered). The diameter is about four miles.

Wirt's aphelion is 3.1994 AU, the perihelion is 1.5617 (crossing the orbit of Mars, and so in the Aten group), and eccentricity .3440. It rotates in 3.69 hours. The orbital period is 1341.60 days, or 3.67 years. The albedo is .1907.

* *S/2005 (2044) 1* is a 1.2 mile diameter moon discovered by Donald Pray and a dozen colleagues. It has a period of 18.97 hours

2062 Aten was discovered January 7, 1976, by Eleanor F. Helin (1932-2009; asteroid 3267 Glo, as Glo was her nickname). This is one of around 800 asteroids credited to Helin; however. when I interviewed her for my cable show in 1996, she said her real count was closer to 6,000. I did not edit out my gasp! The formal count does not consider asteroids whose orbits remain undetermined. Aten was an aspect of the Egyptian sun god. The diameter is about 0.7 mile.

The aphelion is 1.1436 AU, the perihelion 0.7902 AU, with an eccentricity of .1828. The inclination is 18.93 degrees. The orbital period is 347.27 days, or 0.95 year. It rotates in 40.77 hours.

Aten is class S with an albedo of .26. The surface gravity is 0.00003G, with an escape velocity of 2 feet per second.

This is the namesake of the Aten group of over 200 asteroids. They are defined as having an aphelion in the asteroid belt, and a perihelion inside the orbit of Mars.

2074 Shoemaker was discovered October 17, 1974 by Helin. It is named for Gene Shoemaker (1928-1997; a specialist in study of meteor craters).

Shoemaker's aphelion is 1.9469 AU, perihelion of 1.6524 AU, and eccentricity .0818. The orbital inclination is 30.08 degrees. The orbital period is 881.82 days, or 2.41 years. It rotates in 2.533 hours. Geologically it is class S, and is a Mars crosser.

2145 Blaauw was discovered October 24, 1978, by Richard M. West (b. 1941; this is the sixth of 39 asteroids he has found). It is named for Adriaan Blaauw. The diameter is 22 miles.

The aphelion is 3.5266 AU, the perihelion is 2.9109 AU, and eccentricity .09564. The inclination of the orbit to the ecliptic is 15.010 degrees. The orbital period is 2109.26 days, or 5.77 years. It rotates in 12.41 hours.

The albedo is .0869. This is a member of the Jacobus Collective.

2146 Stentor was discovered October 24, 1976, by West. It is a Jupiter Trojan appropriately placed in the Greek (L4, or leading) section.

The aphelion is 5.7260 AU, the perihelion is 4.6655 AU, and the eccentricity is .1021. The orbital inclination is 39.26 degrees. The orbital period is 4325.86 days, or 11.84 years.

2203 van Rhijn was discovered September 28, 1935, by van Gent. It is named for Pieter Johannes van Rhijn.

The aphelion is 3.6889 AU, the perihelion is 2.5316 AU, and eccentricity .1860. The inclination of the orbit is 1.6455 degrees. The orbital period is 2003.51 days, or 5.49 years. It rotates in 30.55 hours. This asteroid is a member of the Jacobus Collective.

2234 Schmadel was discovered April 27, 1977, by Hans-Emil Schuster (b. 1934; #2 of 25 asteroids discovered; 2018 Schuster). It is named for Lutz Schmadel.

Schmadel's aphelion is 3.2417 AU, the perihelion is 2.1573 AU, and the eccentricity .2008. The inclination is 25.23 degrees. The orbital period is 1620.04 days, or 4.44 years. It rotates in 15.63 hours.

2241 Alcathous was discovered November 22, 1979, by Charles Kowal (1940-2011, discovered 19 asteroids, 2 moons of Jupiter, and comets). It is named for a brother-in-law of Aeneas who was said to be the most handsome and brave of the Trojans. The diameter is about 70.4 miles.

Alcathous has an aphelion of 5.5343 U, a perigee of 4.8513 AU, and eccentricity .0658. The orbit is inclined 16.613 degrees. The orbital period is 4322.13 days, or 11.83 years. It rotates in 7.69 hours.

This asteroid has a D spectrum, and an albedo of .0421.

2261 Keeler was discovered April 20, 1977, by Arnold R. Klemola. It is named for James E. Keeler, a former director of Lick Observatory, where it was discovered.

The aphelion is 2.9407 AU, the perihelion 1.8170 AU, and the eccentricity .2362. The orbital inclination is 22.74 degrees. The orbital period is 1340.15 days, or 3.67 years. It rotates in 23.10 hours.

2290 Helffrich was discovered February 14, 1932, by Max Wolf. It is named for Joseph Helffrich (1890-1971; discovered 14 asteroids).

Helffrich's aphelion is 3.1966 AU, the perihelion is 1.9884 AU, and eccentricity .2330. The orbit is inclined 11.55 degrees. The orbital period is 1524.66 days, 4.17 years.

2308 Schilt was discovered May 6, 1967, by Carlos U. Cesco (d. 1987), the ninth of 20 asteroids he discovered. It is named for Jan Schilt, a former student of Kapteyn (818 Kapteynia) and van Rhijn (2203 Vanrhijn). In addition to teaching at Columbia University from 1933 to 1962, Schilt was one of the founders of the observatory where this asteroid was discovered. The diameter is about 10.9 miles.

Schilt has an aphelion of 2.9892 AU, a perihelion of 2.10585 AU, and an eccentricity of .1734. The inclination is 14.18 degrees. Its orbital period is 1485 days, or 4.07 years. It rotates in 9.767 hours.

It is class S with an albedo of .1094. This asteroid is a member of the Jacobus Collective.

2340 Hathor was discovered October 22, 1976, by Kowal. Hathor was an Egyptian goddess often shown as a hippopotamus. The diameter is only 0.2 mile.

Hathor's aphelion is 1.2240 AU, the perihelion 0.4645 AU, and eccentricity .4498. The orbital inclination is 5.854 degrees. Its orbital period is 283.32 days, or 0.78 year, so as an Earth crossing asteroid it is in the Aten class.

2359 Debehogne was discovered October 5, 1931, by Reinmuth. In 2003, it was named for Henri Debehogne (1928-2007).

Debehogne's aphelion is 2.7086 AU, the perihelion is 2.1403 AU, and eccentricity .1172. The orbit is inclined 4.34 degrees to the ecliptic. The orbital period is 1378.86 days, or 3.78 years.

2325 Chernykh was discovered September 25, 1979, by Antonin Mrkos (1832 Mrkos; the third of 274 asteroids he discovered). It is named for Lyudmila Chernykh, who discovered the asteroid named for Mrkos!

This asteroid has an aphelion of 3.6810 AU, a perihelion of 2.6033 AU, and an eccentricity of .1715. The inclination of the orbit is 1.9157 degrees. The orbital period is 2034.44 days, or 5.57 years.

2439 Ulugbek was discovered August 21, 1977, by Nikolai Chernykh (husband of Lyudmila Chernykh). It is named for Ulugh Beigh (1394-1449) a fifteenth century sultan and astronomer. The diameter is 12.6 miles.

Ulugbek has an aphelion of 3.6218 AU, a perihelion of 2.6392 AU, and eccentricity of .1576. The inclination of the orbit is a very low 0.283 degree. The orbital period is 2025.54 days, or 5.55 years.

The albedo is .1065.

2594 Acamas was discovered October 4, 1978, by Kowal. Acamas was a Greek fighter in the Trojan War and was one of those concealed in the Trojan Horse.

The aphelion is 5.4939 AU, the perihelion 4.6454 AU, and the eccentricity .0837. The orbital inclination is 5.5325 degrees. The orbital period is 4169.3 days, or 11.42 years. It is a Jupiter Trojan.

2709 Sagan was discovered March 21, 1982, by Bowell. It is named for Carl Sagan (1934-1996).

Sagan's aphelion is 2.3488 AU, the perihelion is 2.0413 AU, and the eccentricity is .0700. The orbital inclination is 2.733 degrees. The orbital period is 1187.84 days, or 3.25 years. It rotates in 5.26 hours.

This is a class S asteroid.

2732 Witt was discovered March 19, 1926, by Max Wolf. It is named for Carl Gustav Witt (1866-1946).

Witt's aphelion is 2.8272 AU, the perihelion is 2.6910 AU, and the eccentricity .0247. The orbit is inclined 6.494 degrees. The orbital period is 1673.97 days, or 4.58 years.

This is a class A asteroid.

2822 Sacajawea was discovered March 14, 1980, by Bowell. It is named for the woman (1788-1812) who guided the Lewis and Clark expedition.

Sacajawea's aphelion is 2.8998 AU, the perihelion is 2.2647 AU, and eccentricity .1230. The inclination is 14.72 degrees. The orbital period is 1515.66 days, or 4.15 years.

Asteroid Steins September 2008, from several angles. Photo courtesy of ESA.

2867 Steins was discovered November 4, 1969, by N. Chernykh. It is named for Latvian astronomer Karlis Steins (1911-1983). It is 4.2 by 3.6 by 2.8 miles, and is shaped like a diamond in a piece of jewelry.

The aphelion is 2.7064 AU, the perihelion is 2.0202 AU, and the eccentricity is .1452. The orbital inclination is 9.9434 degrees. The orbital period is 1327.01 days, or 3.63 years. It rotates in 6.05 hours.

This class E asteroid is in the Hungaria family. It was photographed from space by the European Space Agency's Rosetta Spacecraft. Craters have been named for precious gemstones, thanks to its shape. One region was named for Chernykh.

2928 Epstein was discovered April 5, 1976, at the Felix Aguilar Observatory. It is named for Isadore Epstein (1919-1995), a Columbia University astronomy professor who helped determine the siting of the Aguilar Observatory (and taught my spherical astronomy class).

The aphelion is 3.1938, the perihelion 2.824 AU, and eccentricity .06145. The orbital inclination is 9.5268 degrees. The orbital period is 1906.38 days, or 5.22 years.

2952 Lilliputia was discovered September 22, 1979, by N. Chernykh. It is named for the land of the little people in **Gulliver's Travels.** The diameter is 5.8 miles.

Lilliputia's aphelion is 2.7081 AU, the perihelion is 1.9188 AU, and eccentricity .1706. The orbital inclination is 3.3215 degrees. The period around the Sun is 1285.24 days, or 3.52 years. It rotates in 3.26 hours.

This class C asteroid has an albedo of .0505.

2975 Spahr was discovered January 8, 1970, by H. Potter and A. Lokalov. It is named for Timothy B. Spahr, head of the Minor Planet Center.

Spahr has an aphelion of 2.4623 AU, a perihelion of 2.0354 AU, and an eccentricity of .0949. The inclination of the orbit to the ecliptic is 6.898 degrees. The orbital period is 1231.81 days, or 3.37 years.

3106 Morabito was discovered March 9, 1981, by Edward Bowell (b. 1943). It is named for Linda Morabito Meyer (b. 1953) who discovered the live volcanoes on Jupiter's moon Io.

The aphelion is 3.8885 AU, the perihelion 2.4028 AU, and eccentricity .2362. The inclination is 14.85 degrees. The orbital period is 2037.82 days, or 5.58 years. It rotates in 6.26 hours.

3116 Goodricke was discovered February 11, 1983, by Bowell. It is named for John Goodricke (1764-1786), a deaf astronomer who was the first to successfully explain the variability of the eclipsing binary star Algol.

The aphelion is 2.6744 AU, the perihelion 1.7804 AU, and the eccentricity is .2007. The inclination of the orbit to the ecliptic is 5.4681 degrees. The orbital period is 1214.20 days, or 3.32 years. It rotates in about 10 hours.

3123 Dunham was discovered August 30, 1981, by Bowell. It is named for David W. Dunham, who first proposed some asteroids may have moons.

Dunham's aphelion is 2.7930 AU, the perihelion is 2.1331 AU, and eccentricity .1340. Inclination to the ecliptic is 1.994 degrees. The orbital period is 1411.90 days, or 3.87 years.

It is class F.

3267 Glo was discovered January 3, 1981, by Edward Bowell. It honors Eleanor Helin, who found thousands of asteroids. The diameter is 8.5 miles.

The aphelion is 3.0173 AU, the perihelion 1.6412 AU (thus approaching Mars's aphelion, and making it a member of the Aten group), and eccentricity of .2954. The inclination is 24.015 degrees. The orbital period is 1298.43 days, or 3.55 years. It rotates in 6.878 hours.

The albedo is .0607.

3310 Patsy was discovered October 9, 1931, by Clyde W. Tombaugh, who named it for one of his grandchildren.

Patsy's aphelion is 3.1891 AU, the perihelion is 2.8299 AU, and eccentricity .0597. The orbital inclination is 11.088 degrees. The orbital period is 1906.97 days, or 5.22 years. It rotates in 9.36 hours.

3500 Kobayashi was discovered September 18, 1919, by Reinmuth. It is named for Japanese astronomer Takeo Kobayashi.

Kobayashi's aphelion is 2.6787 AU, the perihelion is 1.8002 AU, and the eccentricity is .1961. The orbit is inclined 4.256 degrees. The orbital period is 1224.09 days, or 3.35 years.

3582 Cyrano was discovered October 2, 1986, by Paul Wild. It is named for Cyrano de Bergerac (1619-1655) who wrote a couple of science fiction novels about visiting the Moon and is remembered for his large nose and love of Roxanne. There is a well-known opera in which one of its characters is Cyrano de Bergerac, resulting in some people not realizing he was a real person.

Cyrano's aphelion is 3.2394 AU, the perihelion is 2.7636 AU, and eccentricity .0793. The orbital inclination is 10.88 degrees. The orbital period is 1899.36 days, or 5.20 years.

3604 Berkhuijzen was discovered October 17, 1960, by van Houten, Houten-Groeneveld, and Gehrels. It is named for Elly M. Berkhuijzen, a Dutch astronomer specializing in galactic structure.

The aphelion is 2.9029 AU, the perihelion is 2.2974 AU, and the eccentricity .1164. The inclination of the orbit is 12.026 degrees. The orbital period is 1531.40 days, or 4.19 years.

It is a member of the Jacobus Collective.

3673 Levy was discovered on August 22, 1985, by Bowell, one of at least 672 asteroids he has discovered. It is named for David H. Levy, who also has discovered a few asteroids, as well as quite a few comets.

Levy has an aphelion of 2.7779 AU, a perihelion of 1.9188 AU, and an eccentricity of .1842. The inclination of the orbit is 7.0877 degrees. The period is 1312.35 days, or 3.59 years. It rotates in 2.19 hours.

* On December 9, 2007, Don Pray and others discovered Levy has a moon orbiting it in 21.6 hours.

3749 Balam was discovered January 24, 1982, by Bowell. It is named for Canadian astronomer David D. Balam, who has found over 600 asteroids as well as supernovas and comets. The diameter is 4.5 miles.

The aphelion is 2.4813 AU, the perihelion is 1.9930 AU, and the eccentricity .1091. The orbit is inclined 5.38 degrees to the ecliptic. The orbital period is 1222.17 days, or 3.35 years. It rotates in 2.8 hours.

Class S, Balam has an albedo of .16. The density is 2.61.

* On February 13, 2002, William Merline discovered a moon, *S/2002 (3749) 1*, in an extremely elliptical orbit, with eccentricity 0.9, and a semimajor axis of 179.6 miles. The orbital period is 61 days.

* Franck Marchis found a closer and smaller second moon, with a two-mile diameter in March 2008.

3753 Cruithne was discovered October 10, 1986, by Duncan Waldron. The name is that of a possibly mythical Pictish King of Ulster. The diameter is about 3.1 miles.

The aphelion is 1.5113 AU, the perihelion is 0.4841 AU, and the eccentricity is .5148. The orbit's inclination is 19.81 degrees. The orbital period is 363.99 days. Cruithne is the first of five asteroids known to be in a close to 1:1 resonance with Earth. As such, it can have its orbit switched to one with a period slightly longer than Earth's every few centuries. It rotates in 27.44 hours.

The surface gravity is 0.0007G, with an escape velocity of 8.5 feet per second.

3808 Tempel was discovered March 14, 1982, by Boergen. It is named for Ernst W. L. Tempel (1821-1888) who, in addition to several comets, discovered five asteroids.

Tempel's aphelion is 2.6489 AU, the perihelion is 1.9653 AU, and eccentricity .1482. Inclination to the ecliptic is 6.33 degrees. The orbital period is 1279.95 days, or 3.50 years.

3859 Boergen was discovered March 4, 1987, by Bowell. It is named for Freimut Boergen (b. 1930) who has discovered 519 asteroids.

Boergen's aphelion is 3.6413 AU, the perigee is 2.7507 AU, and the eccentricity .1393. The inclination is 2.885 degrees. The orbital period is 2086.97 days, or 5.71 years.

3908 Nyx was discovered August 6, 1980, by H. E. Schuster. It is named for a daughter of Chaos, goddess of the night. The diameter is 0.6 mile.

The aphelion of Nyx is 2.8117 AU, the perihelion 1.0428 AU, and eccentricity is .4589. It crosses the orbit of Mars and comes very close to Earth's aphelion. The orbit is inclined 2.1624 degrees. It rotates in 4.436 hours.

The albedo is .23, with a B spectrum.

3969 Elst was discovered October 16, 1977, by van Houten, van Houten-Groeneveld, and Gehrels. It is named for Belgian astronomer Eric Elst, who also has discovered many asteroids (he even named two for his daughters).

Elst's aphelion is 2.7408 AU, the perihelion is 2.1141 AU, and eccentricity .1291. The inclination of the orbit is 5.645 degrees. The orbital period is 1381.41 days, or 3.78 years. Elst rotates in 6.63 hours.

3992 Wagner was discovered September 29, 1987, by Boergen. It is named for composer Richard Wagner (1813-1883).

Wagner's aphelion is 3.2636 AU, the perihelion is 2.7708 AU, and eccentricity .0817. The inclination is 10.42 degrees. The orbital period is 1914.32 days, or 5.24 years.

4072 Yayoi was discovered October 30, 1981, by H. Kosai. It is named for an archeological era of 300 BC to 300 AD in Japan's past.

Yayoi's aphelion is 2.2841 AU, the perihelion is 2.0061 AU, and eccentricity .0648. The orbital inclination is 2.163 degrees. The orbital period is 1147.54 days, or 3.14 years.

This is an S class asteroid.

4139 Ul'yanin was discovered November 2, 1975, by Tamara Smirnova (asteroid 5540 Smirnova). It is named for airplane designer and developer Sergei A. Ul'yanin (1871-1921).

Ul'yanin's aphelion is 3.6759 AU, the perihelion is 2.6103 AU, and eccentricity .1695. Inclination of the orbit is 1.591 degrees. The orbital period is 2035.34 days, or 5.57 years.

4155 Watanabe was discovered October 25, 1987, by Ueda (asteroid 4676 Uedaseiji) and Kaneda. It is named for Kazuro Watanabe (b. 1955, discovered hundreds of asteroids).

Watanabe's aphelion is 3.0192 AU, the perihelion is 1.8478 AU, and eccentricity .2407. The inclination of the orbit to the ecliptic is 6.0098 degrees. The orbital period is 1386.58 days, or 3.80 years. It rotates in 4.497 hours.

4179 Toutatis was discovered January 4, 1989, by Christian Pollas. Toutatis was a Celtic god roughly equivalent to the Roman Mercury. Its irregular shape is 1.1 by 1.3 by 3 miles.

Toutatis has an aphelion of 4.1297 AU, a perihelion of 0.9373 AU, and an eccentricity of .63003. Its orbital period is 1472.93 days, or 4.03 years. The inclination is 0.4471 degree. It has a slow rotational period of 176 hours.

The albedo of Toutatis is .15. It is class S in the Alinda and Apollo groups. It crosses the orbits of both Mars and Earth. Note the similarity to the orbit of 6489 Golevka.

4244 Zakharchenko was discovered October 7, 1981, by Lyudmila Chernykh. It is named for Ukrainian mathematician Mykhaili Zakharchenko (1825-1912), a specialist in non-Euclidean geometries and textbook author.

Zakharchenko's aphelion is 3.7439 AU, the perihelion is 2.6595 AU, and eccentricity .1693. The orbital inclination is 1.768 degrees. The orbital period is 2092.54 days, or 5.73 years.

4246 Telemann was discovered September 24, 1982, by Boergen. It is named for composer Georg Philipp Telemann (1681-1767).

The aphelion is 2.6034 AU, the perihelion is 1.8301 AU, and eccentricity .1744. The inclination of the orbit to the ecliptic is 3.073 degrees. The orbital period is 1205.52 days, or 3.30 years. Telemann rotates in 8.96 hours.

4279 DeGasparis was discovered November 19, 1982. It is named for Annibale De Gasparis.

The aphelion Is 2.8551 AU, the perihelion is 1.8738 AU, and eccentricity .2075. The orbital inclination is 4.277 degrees. The orbital period is 1327.97 days, or 3.64 years.

4372 Quincy was discovered October 3, 1984, at the Harvard Oak Ridge Observatory and is named for the town in which the observatory is located.

Quincy's aphelion is 3.2935 AU, the perihelion is 2.5709 AU, and the eccentricity .1232. The orbit's inclination is 1.515 degrees. The orbital period is 1833.95 days, or 5.02 years.

This is a class S asteroid.

4435 Holt was discovered January 13, 1983, by Carolyn Shoemaker, who has discovered more than 800 asteroids. It is named for Henry E. Holt (b. 1929, discovered over 600 asteroids).

Holt's aphelion is 3.0943 AU, perihelion of 1.5404 AU, and eccentricity .3353. The orbit is inclined 21.903 degrees to the ecliptic. The orbital period is 1288.51 days, or 3.53 years. It is class S and a Mars crosser.

4492 Debussy was discovered September 17, 1988, by Eric W. Elst (found 3000 asteroids; 3936 Elst). It is named for composer Claude Debussy (1862-1918). The diameter is about 6 miles.

Debussy's aphelion is 3.2626 AU, the perihelion is 2.2714 AU, and the eccentricity is .1791. The inclination of the orbit to the ecliptic is 8.029 degrees. The period is 1681.2 days, or 4.60 years. It rotates in 26.6 hours, synchronous with its moon. The density is only 0.91.

* *S/2004 (4492) 1* was discovered March 21, 2004 by R. Behrend, S. Sposetti and colleagues. It has an elliptical orbit from 3 to 8 miles from Debussy, taking 26.6 hours, and a 4 mile diameter.

4660 Nereus was discovered February 28, 1982, by Eleanor Helin. Nereus was a Titan associated with the sea in Greek mythology and the father of the fifty Nereids. It is only about 1,260 feet long.

Nereus has an aphelion of 2.0246 AU, a perihelion of 0.9537 AU, and an eccentricity of .3600. Its orbital period is 663.43 days, or 1.82 years. The inclination is 1.4319 degrees. It rotates in about 15.1 hours.

This asteroid has an extremely high albedo of .55. It is in the Apollo group.

4676 Uedaseiji was discovered September 16, 1990, by Fujii and Kazuro Watanabe (b. 1955; asteroid 4155 Watanabe). It is named for Seiji Ueda (b. 1952; discovered or co-discovered over 700 asteroids).

The aphelion is 2.5883 AU, the perihelion 2.2143 AU, and eccentricity .0779. The orbital inclination is 8.991 degrees. The orbital period is 1359.12 days, or 3.72 years.

4897 Tomhamilton was discovered August 22, 1987, by Eleanor Helin. It is available for observing sessions and flybys. First landing receives twenty-percent discount on parking fees. The diameter is estimated to be around 35 miles.

The aphelion is 3.4376 AU, the perihelion is 2.6748 AU, and eccentricity .1248. The inclination is 11.065 degrees. The orbital period is 1951.54 days, or 5.34 years.

This asteroid is a member of the Jacobus Collective.

4923 Clarke was discovered March 2, 1981, by Schelte J. Bus. It is named for Arthur C. Clarke (1917-2008), inventor of communication satellites and noted science-fiction author.

Clarke's aphelion is 2.5787 AU, the perihelion is 1.7111 AU, and the eccentricity .2022. Orbital inclination is 6.670 degrees. The orbital period is 1147.37 days, or 3.14 years. This is a class S asteroid.

5020 Asimov was discovered March 2, 1981, by S. Bus. It is named for Isaac Asimov (1920-1992).

Asimov's aphelion is 2.6124 AU, the perihelion is 1.6961 AU, and the eccentricity is .2127. The inclination of the orbit is 1.0999 degrees. The orbital period is 1154.92 days, or 3.16 years.

5145 Pholus was discovered January 9, 1992, by Rabinowitz using Spacewatch data. It is named for a centaur.

The aphelion of Pholus is 31.966 AU, the perihelion is 8.651 AU, and the eccentricity .5740. The orbit is inclined 24.73 degrees. The orbital period is 33428.7 days, or 91.52 years. It rotates in 9.98 hours. This object crosses the orbits of Saturn, Uranus, and Neptune.

The albedo is .044, with a surface believed to have dark amorphous carbon, water ice, methanol ice (CH3OH), olivine, and tholin.

5231 Verne was discovered May 9, 1988, by Carolyn Shoemaker. It is named for science-fiction author Jules Verne (1828-1905).

Verne's aphelion is 3.0165 AU, the perihelion is 2.2226 AU, and eccentricity .1515. The orbital inclination is 14.902 degrees. The orbital period is 1548.57 days, or 4.24 years. It rotates in 4.32 hours.

5261 Eureka was discovered June 28, 1990, by David H. Levy (b. 1948; asteroid 3673 Levy). Levy is best known for co-discovering the comet that collided with Jupiter, among a couple dozen other comets he has found. This was the first of 41 asteroids he has thus far discovered. The diameter is 1.8 miles.

With an aphelion of 1.6221 AU, a perihelion of 1.4250 AU and an eccentricity of .0647, Eureka is the first of four known asteroids in Trojan positions of Mars's orbit. It is the only one at the L5 (trailing) position. Its orbital period is 686.86 days, or 1.88 years. The inclination of the orbit is 20.28 degrees. It is believed to rotate in about six hours.

It is class S, somewhat different than the asteroids at Mars's L4 position.

5335 Damocles was discovered February 18, 1991, by Robert H. McNaught (1956-), one of 469 asteroids he has discovered. The name remembers the Greek philosopher with a sword hanging over him. The diameter is about six miles.

Damocles has an aphelion of 22.094 AU (beyond the orbit of Uranus), a perihelion of 1.5789 AU (near the orbit of Mars), and an eccentricity of .8666. The inclination of the orbit to the ecliptic is 62.01 degrees. The orbital period is 14,874.1 days, or 40.72 years. It rotates in 10.2 hours.

It is believed to be Class S.

Damocles was the first discovered of what are now 81 asteroids forming the Damocloid Family. These asteroids are characterized by orbits with perihelion under 5.2 AU, a semimajor axis greater than

8.0 AU, and an eccentricity greater than .75, but less than 1.0. They have a very low albedo, and a somewhat reddish tint. It is believed most are the dormant nuclei of comets which have lost all their volatiles.

5535 Annefrank was discovered March 23, 1942, by Reinmuth. It is named for the hidden World War II diarist Anne Frank (1929-1945). It is 3.8 by 3.1 by 2 miles.

Annefrank's aphelion is 2.3548 AU, the perihelion is 2.0713 AU, and eccentricity .0641. Inclination of the orbit is 4.2470 degrees. The orbital period is 1202.49 days, or 3.29 years. It rotates in 15.12 hours.

The albedo is .21, with a spectral class of S.

Although this asteroid was visited by a spacecraft and photographed, all the photos returned were so blurred and out of focus that none was worth reproducing here.

5540 Smirnova was discovered August 30, 1971, by Tamara Smirnova (1935-2001), for whom it is named.

Smirnova's aphelion is 3.3993 AU, the perihelion is 1.7973 AU, and eccentricity .3083. The orbit is inclined 4.559 degrees. The orbital period is 1529.81 days, or 4.19 years.

5655 Barney was discovered September 29, 1973, by van Houten, van Houten-Groenwald, and Gehrels. It is named for astronomer Ida Barney (1886-1982) who compiled an enormous data base in the field of astrometrics, and battled sexism in the profession. Some of her work was done in collaboration with Jan Schilt (asteroid 2308).

Barney's aphelion is 2.6788 AU, the perihelion is 2.4781 AU, and eccentricity .0389. The inclination of the orbit is 14.497 degrees. The orbital period is 1512.28 days, or 4.14 years.

6090 1989DJ was discovered February 27, 1989, by Henri Debehogne (1928-2007). The diameter is 46.3 miles.

The aphelion is 5.6202 AU, the perihelion 5.0107 AU, and eccentricity .0573. The inclination is 20.18 degrees. The orbital period is 4476.15 days, or 12.26 years. It is a Jupiter Trojan. It rotates in 18.48 hours.

The albedo is .0553.

6312 Robheinlein was discovered September 14, 1990, by Henry E. Holt. It is named for science fiction writer Robert A. Heinlein (1907-1988).

Robheinlein's aphelion is 2.3355 AU, the perihelion is 2.0311 AU, and eccentricity .0697. The inclination of the orbit to the ecliptic is 4.114 degrees. The orbital period is 1178.33 days, or 3.23 years.

6469 Armstrong was discovered by Mrkos on August 14, 1982. It is named for Neil Armstrong (1930-2012), first person to set foot on the Moon, July 20, 1969.

The aphelion is 2.6716 AU, the perihelion 1.7702 AU, and the eccentricity is .20293. The orbital inclination is 3.955 degrees. The orbital period is 1208.91 days, or 3.31 years.

6470 Aldrin was discovered September 14, 1982, by Mrkos. It is named for Edwin "Buzz" Aldrin (b. 1930), second man on the Moon.

Aldrin's aphelion is 2.6199 AU, the perihelion is 1.9305 AU, and the eccentricity .1515. The inclination of the orbit is 2.7917 degrees. The orbital period is 1253.53 days, or 3.43 years.

6471 Collins was discovered March 4, 1983, by Mrkos. It is named for Michael Collins (b. 1930), command pilot on the first manned-landing on the Moon.

Collins's aphelion is 2.7321 AU, the perihelion is 2.1314 AU, and the eccentricity .1237. Orbital inclination is 2.674 degrees. The orbital period is 1335.52 days, or 3.79 years.

6489 Golevka was discovered May 10, 1991, by Helin. The name combines the first letters of three observatories that studied it with radar: Goldstone, California; Yevpatoria, Ukraine; and Kashima, Japan. It is .4 by .85 mile.

Golevka's aphelion is 4.0153 AU, the perihelion is 0.9749 AU, and the eccentricity .6093. The inclination is 2.271 degrees. The orbital period is 1439.57 days, or 3.94 years. It rotates in 6.026 hours.

The albedo is .15, and the density 2.7.

The orbit puts it into the Apollo and Alinda groups and is strikingly similar to the orbit of 4179 Toutatis. Golevka was the first asteroid ever shown to experience the Yarkovsky effect, thanks to the radar studies.

6639 Marchis was discovered September 25, 1989, by Debehogne. It is named for Franck Marchis (b. 1973).

Marchis has an aphelion of 3.5881 AU, a perihelion of 2.7472 AU, and an eccentricity of .1327. Inclination to the ecliptic is 2.537 degrees.

Orbital period is 2059.20 days, or 5.64 years.

7066 Nessus was discovered April 26, 1993, by Rabinowitz using Spacewatch data. Nessus was one of the centaurs in Greek legends.

The aphelion of Nessus is 37.163 AU, the perihelion is 11.781 AU, and eccentricity .5186. The orbital inclination is 15.65 degrees. The orbital period is 44,218.55 days, or 121.06 years.

7100 Martin Luther was discovered September 29, 1973 by van Houten, van Houten-Groeneveld, and Gehrels. It is named for the German religious leader of the Reformation.

The aphelion is 3.1185 AU, the perihelion is 2.6196 AU, and eccentricity .0869. The inclination is 1.2650 degrees. The orbital period is 1775.01 days, or 4.86 years.

*7211 Xerxes w*as discovered March 25, 1971, by van Houten, van Houten-Groeneveld, and Gehrels. It is named for Xerxes I (519-465 B.C.), ruler of the Persian Empire.

Xerxes has an aphelion of 3.2574 AU, a perihelion of 2.3377 AU, and eccentricity .1644. The orbital inclination is 9.1205 degrees. The orbital period is 1709.12 days, or 4.68 years.

This is a class S asteroid.

7231 Porco was discovered October 15, 1985, by Bowell. It is named for the prominent planetary scientist Carolyn Porco (b. 1953; discovered three moons of Saturn, including Daphnis and Aegaeon).

Porco's aphelion is 3.4055 AU, the perihelion is 2.9330 AU, and eccentricity .0745. The inclination of the orbit is 9.426 degrees. The orbital period is 2060.77 days, or 5.64 years.

7550 Woolum was discovered March 1, 1981, by Schelte Bus. It is named for Dorothy S. Woolum (b. 1942), Professor Emerita of astrophysics at California State University at Fullerton.

Woolum's aphelion is 2.7364 AU, the perihelion is 1.8445 AU, and eccentricity .1917. The orbit is inclined 5.527 degrees to the ecliptic. The orbital period is 1266.16 days, or 3.47 years. It rotates in 3.47 hours.

7907 Erasmus was discovered September 24, 1960, by van Houten, van Houten-Groeneveld, and Gehrels. It is named for philosopher Erasmus of Rotterdam (1469-1536).

The aphelion of Erasmus is 2.8384 AU, the perihelion is 2.7854 AU, and eccentricity .0190. The orbital inclination is 2.0677 degrees. It revolves around the Sun in 1698.01 days, or 4.65 years.

8191 Mersenne was discovered July 20, 1993, by Eric Elst. It is named for Merin Mersenne (1588-1648; 8191 is the fifth Mersenne prime).

Mersenne's aphelion is 2.5117 AU, the perihelion is 2.0157 AU, and the eccentricity .1096. The inclination of the orbit is 2.8283 degrees. The orbital period is 1244.03 days, or 3.41 years.

8558 Hack was discovered August 1, 1995, by Luciano Tesi (b. 1931, fourth of 175 asteroids he has discovered). It is named for the prominent Italian stellar astronomer Margherita Hack (1922-2013).

Hack's aphelion is 3.7855 AU, the perihelion is 2.4713 AU, and eccentricity .210. The orbital inclination is 0.286 degree. The orbital period is 2021.05 days, or 5.53 years.

9252 Goddard was discovered October 17, 1960, by van Houten, van Houten-Groeneveld, and Gehrels. It honors Robert H. Goddard (1882-1945), father of the American space program, who had to fight Porlockian doubters from the New York *Times* to supposed colleagues, to start the USA on the pathway to space. His widow had to sue to win millions in royalties for patents on liquid fueled rockets from a government that had tried to ignore him.

The aphelion is 3.1653 AU, the perihelion 2.5742 AU, and the eccentricity is .1682. The orbital inclination is 3.1932 degrees. The orbital period is 1988.53 days, or 5.44 years.

9253 Oberth was discovered March 25, 1971, by van Houten, van Houten-Groeneveld, and Gehrels. It is named for Hermann Oberth (1894-1989, a major figure in the development of space travel, even after his Ph.D. dissertation on it was rejected as utopian).

The aphelion is 2.8316 AU, the perihelion is 1.9580 AU, and eccentricity .1824. The orbital inclination is 7.2776 degrees. The orbital period is 1353.65 days, or 3.71 years.

9531 Jean-Luc was discovered August 30, 1981 by Bowell. It is named for Jean-Luc Margot (b. 1969).

Jean-Luc's aphelion is 2.6510 AU, the perihelion is 2.2340 AU, and the eccentricity .1867. The inclination to the ecliptic is 1.817 degrees.

It orbits the Sun in 1219.59 days, or 3.34 years.

9762 Hermannhesse was discovered September 13, 1991, by Boerngen and Schmadel. It is named for Hermann Hesse (1877-1962), who won the Nobel Prize for Literature in 1946 and is noted for playing games with beads.

The aphelion is 2.3930 AU, the perihelion is 2.1020, and eccentricity .0647. The inclination is 3.7288 degrees. The orbital period is 1230.70 days, or 3.37 years.

9861 Jahreiss was discovered September 9, 1991, by Schmadel (2234 Schmadel) and Boerngen. It is named for Hartmut Jahreiss, who worked with Gliese to make an intensive study of nearby stars (see my book *Our Neighbor Stars*, 2012, for details).

The aphelion is 2.5485 AU, the perihelion is 1.9022 AU, and the eccentricity .1452. The orbit is inclined 2.929 degrees to the ecliptic. The orbital period is 1212.55 days, or 3.32 years.

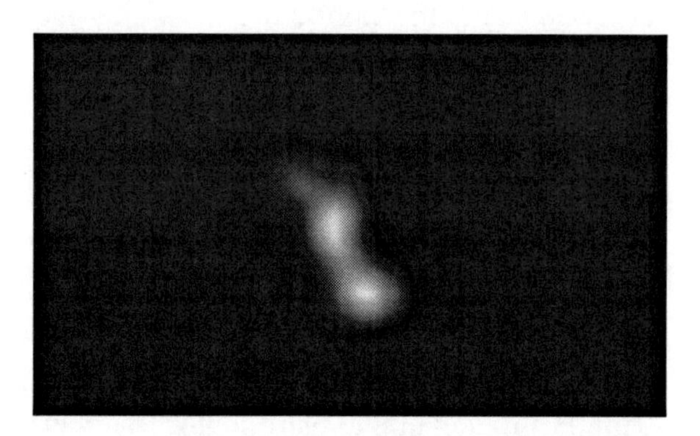

9969 Braille from the Deep Space 1 Spacecraft about 15 minutes after closest approach. Note that despite the poor quality it appears Braille is the result of two objects merging.

9969 Braille was discovered May 27, 1992, by Helin and K. J. Lawrence. It is named for Louis Braille (1809-1852). It is 1.3 by 0.6 by 0.6 miles.

Braille's aphelion is 3.3552 AU, the perihelion is 1.3276 AU (crossing Mars's orbit), and eccentricity .4330. The orbital inclination is 28.99 degrees. The orbital period is 1308.64 days, or 3.58 years. It rotates in the remarkably slow 226.4 hours.

The spectrum is the rare class Q, with olivine and pyroxene on the surface.

10000 Myriotos was discovered September 30, 1951, by Albert Wilson. The name means *ten thousand* in Greek. It has a diameter of 1.8 miles.

Myriotos has an aphelion of 3.3720 AU, a perihelion of 1.8043 AU, and eccentricity of .3029. The inclination is 20.61 degrees. The orbital period is 1520.84 days or 4.16 years.

10199 Chariklo was discovered by Felipe Braga-Ribas at the National Observatory of Brazil. It is named for a centaur married to the leader of centaurs, Chiron.

The aphelion is 18.458 AU, the perihelion is 13.051 AU, and eccentricity .1716. This gives it a period of 62.53 years. The inclination is a very high 23.412 degrees. The diameter is 155 miles.

On June 3, 2013 in a stellar occultation this asteroid was discovered to have two rings, a denser one with a width of four miles and a more distant two mile wide ring.

10430 Martschmidt was discovered September 24, 1960, by van Houten, van Houten-Groeneveld, and Gehrels . It is named for Maartin Schmidt.

The aphelion is 3.0002 AU, perihelion is 2.4827 AU, and eccentricity is .09438. The orbital inclination is 2.741 degrees. The orbital period is 1657.96 days, or 4.54 years.

This asteroid is a member of the Jacobus Collective.

11100 Lai was discovered May 22, 1995, by Luciano Lai (b. 1948).

Lai's aphelion is 2.8431 AU, the perihelion 2.0151 AU, and the eccentricity .1709. The inclination of the orbit is 10.176 degrees. The orbital period is 1382.79 days, or 3.79 years. It rotates in about 5 hours.

12050 Humecronyn was discovered April 27, 1997, by Spacewatch. It is named for Canadian politician Hume B. Cronyn, Sr. (1864-1933), who was instrumental in the founding of Canada's national research arm and was the father of the well-known actor of similar name.

The aphelion is 3.1658 AU, the perihelion is 2.6624 AU, and eccentricity .0864. The orbital inclination is 1.2329 degrees. The orbital period is 1817.00 days, or 4.97 years.

12101 Trujillo was discovered May 1, 1998, by the Lowell Observatory. It is named for Chadwick A. Trujillo (b. 1973), an astronomer active in the study of asteroids.

Trujillo's aphelion is 3.2697 AU, the perihelion is 2.7012 AU, and eccentricity .0952. The inclination of the orbit is 10.26 degrees. The orbital period is 1884.12 days, or 5.16 years.

12279 Laon was discovered November 16, 1990, by Elst. It is named for a town in France, the capitol of the department of Aisne.

Laon's aphelion is 3.0254 AU, the perihelion is 2.5190 AU, and eccentricity .0923. The orbital inclination is 10.269 degrees. The orbital period is 1633.62 days, or 4.61 years.

12927 Pinocchio was discovered September 30, 1999, by Tesi. It is named for Jiminy Cricket's friend.

Pinocchio's aphelion is 2.6148 AU, the perihelion is 1.9936 AU, and the eccentricity .1348. The orbital inclination is 3.809 degrees. Orbital period is 1277.54 days, or 3.50 years.

13092 Schroedinger was discovered September 24, 1992, by Freimut Boerngen and Lutz Schmadel. Erwin Schroedinger (1887-1961) won the 1933 Nobel Prize in physics and was noted for ambiguous treatment of cats.

Schroedinger's aphelion is 2.0619 AU, the perihelion is 2.0460 AU, and eccentricity is .0501. The inclination of the orbit is 0.6172 degree. The orbital period is 1154.64 days, or 3.16 years.

13421 Holvorcem was discovered November 11, 1999, by Charles W. Juels (1944-2009; 473 asteroids discovered). It is named for Brazilian astronomer Paulo R. Holvorcem (190 asteroids discovered).

Holvorcem's aphelion is 2.9119 AU, the perihelion is 2.3983 AU, and the eccentricity .0967. The orbit is inclined 3.294 degrees. The orbital period is 1580.22 days, or 4.33 years.

14120 Espenak was discovered August 27, 1998, by the Near Earth Object program at Lowell Observatory. It is named for Fred Espenak, NASA's long-time expert on eclipses and author of the definitive modern books on eclipses, succeeding von Oppolzer. The diameter is 8.2 miles.

The aphelion is 2.5937 AU, perihelion 2.1731 AU, and eccentricity is .0882. The inclination of the orbit is 5.991 degrees. The period is 1343.96 days, or 3.68 years.

The albedo is .0882. It is a member of the Jacobus Collective.

15151 Wilmacherup was discovered March 4, 2000, by the Catalina Sky Survey. It is named for Wilma Cherup (1915-2010) executive secretary of the Astronomical League 1954-1977.

The aphelion is 2.9074 AU, the perihelion is 1.8140 AU, and eccentricity .2316. The inclination of the orbit is 2.228 degrees. The orbital period is 1324.81 days, or 3.63 years. It rotates in 4.56 hours.

15810 1999 JR1 was discovered May 2, 1994, by M. J. Irwin and A. Zytkow. The diameter is about 68 miles.

The aphelion is 43.720 AU, the perihelion is 34.73 AU, and eccentricity .1146. Orbital inclination is 3.806 degrees. This leaves it in a 2:3 resonance with Neptune and a temporary co-orbital with Pluto. The orbital period is 89728.3 days, or 245.66 years.

The albedo is .09.

16069 Marshafolger was discovered September 7, 1999, by LINEAR. It is named for Marsha Folger of Connecticut, who mentored a finalist in the 2003 Intel Science Talent Search.

Marshafolger's aphelion is 2.9949 AU, the perihelion is 2.0764 AU, and eccentricity is .1832. The inclination of the orbit is 5.333 degrees. The orbital period is 1470.88 days, or 4.03 years.

17459 Andreashofer was discovered October 13, 1990, by Freimut Boerngen (1930-; discovered over 500 asteroids; asteroid 3859 Boerngen) and Lutz Schmadel (1942-; discovered many asteroids, usually with Boerngen; asteroid 2234 Schmadel). It is named for Austrian national hero Andreas Hofer (1767-1810), also a fourth great-grandfather of another asteroid's honoree.

Andreashofer's aphelion is 2.6619 AU, the perihelion is 1.8439, and the eccentricity .1815. The orbital period is 1235.10 days, or 3.38 years.

17601 Sheldonschafer was discovered September 19, 1995, by Timothy Spahr (asteroid 2975 Spahr; he is best known for discovering Jupiter's moon Callirrhoe). Schafer is planetarium director at the Peoria Riverfront Museum and college professor at Bradley University who built the world's largest scale model of the Solar System.

The aphelion is 1.9368 AU, perihelion is 1.7027 AU, and eccentricity is .0643. The orbital inclination is 24.20 degrees. The orbital period is 896.62 days, or 2.45 years.

It is a member of the Jacobus Collective.

23405 Nisyros was discovered September 19, 1973, by van Houten, van Houten-Groeneveld and Gehrels. It is named for a volcano in Greece.

The aphelion is 4.4645 AU, the perihelion 3.4341 AU, and the eccentricity is .1304. The orbit is inclined 5.2175 degrees. The period is 2866.70 days, or 7.85 years.

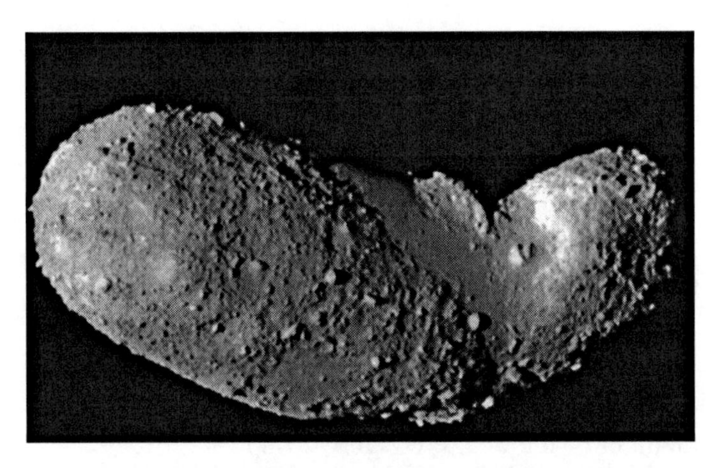

Itokawa from 8 miles away showing the curiously crater-free central region and craters at opposite ends. Photo courtesy of JAXA.

25143 Itokawa was discovered September 26, 1998, by the LINEAR program. The name honors Hideo Itokawa (1912-1999), father of the Japanese space program. It is 3325 by 2000 by 1300 feet, with a notably crater-free region that has a number of boulders showing. It was the first asteroid from which samples were taken and returned to Earth.

The aphelion is 1.6951 AU (just beyond Mars's aphelion), the perihelion is 0.9531 (just inside the orbit of Earth, putting it into the Apollo group), and eccentricity .28018. The inclination is 1.6216 degrees. The orbital period is 556.52 days, or 1.52 years. It rotates in 12.13 hours.

This is an S class asteroid, with a density of 1.92 and an albedo of .53. The surface gravity is 0.00002G (so low the spacecraft that visited could not orbit it), and escape velocity is 8 *inches* per second.

28220 York was discovered December 28, 1998, by J. Ticha and M. Tichy. It is named for the attractive and historic city in England (which also has a terrestrial namesake).

York's aphelion is 2.6927 AU, perihelion is 2.3514 AU, and eccentricity is .0677. The inclination of the orbit is 1.273 degrees. The orbital period is 1462.93 days, or 4.01 years.

28978 Ixion was discovered May 22, 2001, by the Deep Ecliptic Survey. It is named for the ancestor of all centaurs in Greek mythology. The diameter is 400 miles.

Ixion's aphelion is 49.008 AU, the perihelion is 29.742 AU, and the eccentricity is .2446. The orbital inclination is 19.632 degrees. The orbital period is 90247.44 days, or 247.08 years.

The surface seems to be a mix of water ice, carbon, and tholins.

2001 QR322 was discovered August 21, 2001, by the Deep Ecliptic Survey. It has a diameter of approximately 65 miles.

The aphelion is 31.711 AU, the perihelion is 29.470 AU, and eccentricity .0280. The inclination of the orbit is 1.321 degrees. The orbital period is 60983 days, or 166.96 years.

This object is the first Neptune Trojan to be discovered, at the L4 position.

37556 Svyaztie was discovered August 28, 1982, by Nikolai Chernykh (1931-2004) and Brian Marsden. The name combines Russian and English words meaning *connection.*

The aphelion is 2.8337 AU, the perihelion 1.7590 AU, and eccentricity .2340. The orbital inclination is 5.4086 degrees. The period is 1271.02 days, or 3.48 years.

42355 Typhon was discovered February 5, 2002, by the NEAT project at Mount Palomar. It is named for a monster in Greek mythology. The diameter is about 93 miles.

Typhon's aphelion is 58.3498 AU, the perihelion is 17.5145 AU (crossing Neptune's orbit), and the eccentricity is .5383. The orbit is inclined 2.426 degrees. The orbital period is 85,331.44 days, or 233.62 years. It rotates in roughly 5 hours.

This orbit makes it a member of the Scattered Disk group.

* A moon, *Echidna,* was discovered January 20, 2006, by K. Noll and W. Grundy using the Hubble Space Telescope. They named it for Typhon's mate, the mother of nearly all Greek monsters. It has a diameter of about 42 miles, and orbits Typhon at a distance of 800 miles in 11 days.

50000 **Quaoar** *w*as discovered June 5, 2002, by Trujillo and Brown. It is named for a creator god of a Native American people. The diameter is about 690 miles.

Quaoar's aphelion is 45.116 AU, the perihelion is 41.695 AU, and eccentricity .0394. The orbit's inclination is 7.996 degrees. It rotates in 17.67 hours. The orbital period is 104,451 days, or 285.97 years.

The albedo is about .2, with a density of 2.8. The surface gravity is 0.004G, with an escape velocity of about 2000 feet per second.

* *Weywot* was discovered February 22, 2007 by Michael Brown. It is about 9000 miles from Quaoar with an orbital period of 12.438 days. The orbital inclination is 14 degrees. The diameter is believed to be about 46 miles. Weywot was a son of Quaoar.

51865 was discovered August 3, 2001, by the NEAT project.

The aphelion is 4.2676 AU, the perihelion is 3.5777 AU, and the eccentricity is .0879. The inclination of the orbit is 2.5436. The orbital period is 2837.75 days, or 7.77 years.

54509 YORP was discovered August 3, 2000, by LINEAR. The name honors Ivan Yarkovsky, John A. O'Keefe, V. V. Radzievski, and Stephen J. Paddack. The name was selected because it was the first asteroid discovered to show a strong *Yarkovsky* effect, which will result in its destruction in a few thousand years.

YORP's aphelion is 1.2375 AU, the perihelion is 0.7746 AU, and eccentricity .2300. The orbital inclination is 1.5998 degrees. The orbital period is 368.57 days, or 1.01 years. It currently rotates in 12.174 minutes, which the Yarkovsky effect will increase until it is torn apart.

58534 Logos was discovered February 4, 1997, by Chadwick Trujillo (b. 1973; discoverer of many Trans-Neptunian asteroids; 12101 Trujillo) and colleagues. It is named for a creator concept in Gnosticism, with a complementary element named Zoe. The diameter is 48 miles.

The aphelion is 50.8745 AU, the perihelion 39.7362 AU, and the eccentricity .1229. The inclination is 2.897 degrees. The orbital period is 111,384.08 days, or 304.9 years.

The density is 1.0, with a surface gravity of 0.007G, and an escape velocity of 100 feet per second. Logos has an albedo of .39.

* *Zoe*, S/2001 (58534) 1 was discovered November 17, 2001, by the Hubble Space Telescope. It has a diameter of 46 miles, and an eccentricity of .546 with a semimajor axis of 5106 miles. The orbital period is 309.9 days. The name marks a reciprocal concept to Logos in Gnosticism.

64070 NEAT was discovered September 24, 2001, by Charles Juels and Paulo Holvercem. It is named for the Near Earth Asteroid Tracking program.

NEAT's aphelion is 2.9787 AU, the perihelion is 1.9493 AU, and eccentricity .2089. The obit is inclined 12.504 degrees. The orbital period is 1412.74 days, or 3.90 years.

90377 Sedna was discovered November 14, 2003, by Brown, Trujillo and Rabinowitz. It is named for an Inuit sea goddess. The diameter is about 600 miles.

Sedna's aphelion is 937 AU, the perihelion 76.361 AU, and eccentricity .8527. The orbital period is approximately 11,400 years—with such an extreme orbit, precise figures are hard to obtain. It rotates in about 10 hours.

The surface is believed to have water ice, methane, nitrogen ice, and tholins. The surface gravity is about 0.0026G, with an escape velocity of 1,700 feet per second.

90482 Orcus was discovered February 17, 2004, by Brown, Trujillo, and Rabinowitz. It is named for the Etruscan god of the dead. The diameter is about 570 miles.

The aphelion of Orcus is 48.07 AU, the perihelion is 30.27 AU, and eccentricity .2272. The orbital inclination is 20.573 degrees. The orbital period is 89,552 days, or 245.18 years. This gives it a 2:3 resonance with Neptune. It rotates in 13.188 hours.

The albedo is .28, with a density of 2.3. The surface gravity is 0.048G, and escape velocity 1470 feet per second.

* *Vanth* was discovered November 13, 2005, by Michael Brown and T.-A. Suer. It orbits 5600 miles from Orcus in 9.54 days. The inclination is 21 degrees, and rotation is synchronous. The albedo is .12, with a color far more reddish than Orcus, making it most likely a captured moon. It is named for an Etruscan goddess who guided the souls of the dead to the underworld.

99942 Apophis was discovered June 19, 2004, by Tucker and Tholen. It has a diameter of 1060 feet. It bears the name of an Egyptian god of doom, which has unfortunately encouraged various hysterics and nutcases to fabulate nonsense threats from it.

The aphelion is 1.0985 AU, perihelion is 0.7461, and eccentricity is .1911. Those numbers will be changed considerably after it makes a close pass of Earth in 2029. The inclination is 3.331 degrees. The period is 323.515 days, or 0.89 year. It rotates in 30.4 hours.

It is Class B, with an albedo of .23. Until 2029 it is classed as an Aten, after that its new orbit will make it a member of the Amor group.

101955 Bennu was discovered September 11, 1999, by LINEAR, the Lincoln Laboratory Research Team. It is named for an Egyptian creator god, on the suggestion of a school child in a naming contest. Bennu is intended as the target of the Osiris-Rex mission planned for launch in 2016, arrival at the asteroid in 2019, and returning a sample to Earth in 2023. Astronomers are among the most patient of all scientists. Bennu is believed to have a diameter of 1600 feet.

The aphelion is 1.3554 AU, the perihelion 0.8966 (Apollo group), and eccentricity .2037. The inclination is 6.0349 degrees. The orbital period is 436.42 days, or 1.19 years. It rotates in 4.288 hours.

121514, 1999 UJ7 is all this asteroid has for a name. It was the first Martian Trojan to be discovered at Mars's L4 position on October 30, 1999, by the LINEAR program. The diameter is estimated at no more than 0.6 mile.

The aphelion is 1.5844 AU, perihelion of 1.4646 AU, and eccentricity of .0393. The inclination is 16.75 degrees. The period is 687.58 days.

The spectral class appears to be a mixture of M and P.

134340 **Pluto** was discovered February 18, 1930, by Clyde Tombaugh. It is named for the Roman god of the underworld. The diameter is about 1,710 miles.

Pluto's aphelion is 48.871 AU, the perihelion is 29.697 AU, and the eccentricity .2447. The orbital inclination is 17.152 degrees. The orbital period is 90,465 days, or 247.68 years. It rotates in 6 days 9 hours 17 minutes 36.7 seconds, synchronous with its largest and closest moon, Charon.

The density is 2.03, leading to the belief that it has a rocky core overlaid by a thick layer of ices. Near perihelion, a very thin atmosphere of nitrogen (N2), methane (CH4), and carbon monoxide was detected. The surface gravity is 0.057G, and the escape velocity is 4030 feet per second. The orbit is in a 2:3 resonance with that of Neptune.

* *Charon* was discovered June 28, 1978, by James Christy, 12,160 miles from Pluto, but only 10,290 miles from their mutual barycenter. The surface appears to be water ice and ammonia hydrates. The density is 1.68, with an albedo of .38. The equator appears brighter than either pole, with the south polar cap especially dark. Charon was the boatman who carried souls of the dead to Pluto's realm.

* *Styx* was discovered July 10, 2012, by Mark Showalter, S. Alan Stern, Marc Bluie, Andrew Steffl, and Max Mutchler. The diameter is between 6 and 15 miles. It is roughly 26,000 miles from Pluto, and orbits in 20.2 days. Styx was the name of the river Charon boated over to reach Pluto's realm.

* *Nix* was discovered June 4, 2005, by Hal Weaver, S. Alan Stern, Max Mutchler, Marc Bluie, William Merline, John Spencer, Eliot Young, and Leslie Young. It is about 30,250 miles from Pluto, taking 24.856 days to orbit. The diameter is estimated to be around 30 to 40 miles. Nyx was the mother of Charon, but the alternative spelling was chosen to avoid confusion with asteroid 3908 Nyx.

* *Kerberus* was discovered June 28, 2011, by Showalter. The diameter is estimated at about 18 miles. At 36,600 miles from Pluto it takes 32.1 days to orbit. Kerberus was the three-headed dog guarding the gates to Pluto's realm.

* *Hydra* was discovered at the same time as Nix. It has an estimated diameter of 60 miles, making it the second largest of Pluto's moons. It is 40,200 miles from Pluto, with an orbital period of 38.206 days. Hydra was the serpent that wrestled with Hercules, which has nothing to do with Pluto, but the name was selected with Nix to honor with the moons' initials the New Horizons spacecraft headed for Pluto.

136108 **Haumea** was discovered December 28, 2004, by Michael Brown. It is named for the Hawaiian goddess of childbirth.

Haumea's aphelion is 51.544 AU, the perihelion is 34.721 AU, and eccentricity .195. The orbital inclination is 28.22 degrees. The orbital period is 103,468 days, or 283.28 years. It rotates in 3.9155 hours.

The albedo is a very high .7, with a density of 2.9. The surface gravity is 0.0047G, with an escape velocity of 2750 feet per second.

* Namaka was discovered June 30, 2005 by Brown, Trujillo, and Rabinowitz. It orbits about 15,900 miles from Haumea in 18.2783 days at an inclination of 113.013 degrees. The diameter is about 190 miles. Namaka was a daughter of Haumea.

* Hi'iaka was discovered January 25, 2005, by the same trio that found Namaka. It is about 30,975 miles from Haumea, with a diameter of about 217 miles. The orbital period is 49.12 days at an inclination of 126.356 degrees. Density is about 1.0, with a surface mostly of water ice. Hi'iaka was the patron goddess of the Big Island of Hawaii.

136199 **Eris** was discovered January 5, 2005, by Brown, Trujillo, and Rabinowitz. It is named for the goddess of discord (for example, she's the one who tossed the apple labeled "For the Fairest" which is blamed for the Trojan War). The diameter is about 1,720 miles.

The aphelion of Eris is 97.661 AU, the perihelion 38.255 AU, and eccentricity .4371. The inclination of the orbit to the ecliptic is 44.19 degrees. The orbital period is 204,624 days, or 560.23 years.

The surface gravity is 0.009G, with an escape velocity of 4,540 feet per second, both indicating a density of 2.52.

* *Dysnomia* was discovered September 10, 2005, by Antonine Bouchez and the people who discovered Eris. It has a diameter of about 300 miles, and orbits Eris in 15.774 days at a distance of 20,810 miles. The eccentricity is .01, and the inclination of the orbit to the orbit of Eris is 142 degrees (retrograde).

136472 **Makemake** was discovered March 31, 2005 by Brown, Trujillo, and Rabinowitz. The name is that of the Easter Island god of creation and fertility. The diameter is 900 miles.

Makemake's aphelion is 53.074 AU, the perihelion is 38.509 AU, and eccentricity .159. Orbital inclination is 28.96 degrees. The orbital period is 113,183 days, or 309.88 years.

The density is 1.7 with a very high albedo of .77. The surface gravity is 0.004G, with an escape velocity of 1300 feet per second.

154660 Kavelaars was discovered March 29, 2004 by Balam. It is named for John Kavelaars, who was a co-discoverer of eleven moons of Saturn, eight of Uranus, and four of Neptune.

Kavelaars has an aphelion of 2.1326 AU, a perihelion of 1.7087, and an eccentricity of .1103. The orbit is inclined 22.51 degrees to the ecliptic. The orbital period is 972.25 days, or 2.66 years.

*162173, 1999 JU3, w*as discovered by LINEAR, and a joint NASA, MIT, and Air Force project to find Near Earth Asteroids. It was discovered May 10, 1999. The diameter (a somewhat meaningless concept in so irregularly shaped an object) is 3,315 feet.

This asteroid has an aphelion of 1.4159 AU, a perihelion of 0.9632 AU (thus crossing the Earth's orbit and putting it in the Apollo family), and an eccentricity of .19027. The inclination is 5.8840 degrees.

The period is 473.882 days, or 1.30 years. It is a class C object.

162359 was discovered by Spacewatch on December 27, 1999, the last asteroid discovered in the 1900s.

162359 has an aphelion of 3.2098 AU, a perihelion of 2.1738 AU, and an eccentricity of .19242. The inclination is 1.2941 degrees. The period is 1613.10 days, or 4.42 years.*162360* was discovered by LINEAR on January 2, 2000, the first asteroid of the 2000s.

162360 has an aphelion of 3.0419 AU, a perihelion of 2.2515 AU, and an eccentricity of .1493. The inclination is 11.2145 degrees. The period is 1572.71 days, or 4.31 years.

163249, 2002 GT, was discovered April 3, 2002, by Spacewatch run from the University of Arizona, headed by Tom Gehrels (1925-2011) and Robert McMillan.

The aphelion is 1.7947 AU, the perihelion is 0.89436 AU, with an eccentricity of .33482. The inclination is 6.9685 degrees. The orbital period is 569.456 days, or 1.56 years.

The orbit crossings put it into the Apollo group.

164207, 2004 GU9 was discovered April 13, 2004 by LINEAR. It is noted for having an orbit closely matching Earth's, but it is not a Trojan.

This asteroid has an aphelion of 1.13767 AU, perihelion of 0.86471 AU, and an eccentricity of .13632. The inclination is 13.65 degrees. The orbital period is 365.91 days.

2004 GU9 falls into the Aten group.

285263, 1998 QE2 was discovered August 19, 1998, by LINEAR. The diameter is 1.9 miles.

The aphelion is 3.8043 AU, the perihelion is 1.0387 AU, and the eccentricity is .5710. The inclination of the orbit is 12.854 degrees. The period is 1376.35 days, or 3.77 years. It rotates in about 5 hours.

This is an Amor group member, with an apparent magnitude ranging from 11 on the rare close approaches to Earth (3.6 million miles on May 31, 2013) to 23.9.

* A moon was discovered during the 2013 pass of Earth, first noted by Marina Brozovic. It has a diameter of 2,500 feet, and, although no more than four miles from the primary, takes 32 hours to orbit. Interestingly, it has synchronous rotation.

2006 RH120 was discovered by the Catalina Sky Survey on September 16, 2006. The diameter is about 17 feet.

The aphelion is 1.02004 AU, perihelion 0.97891 AU, with an eccentricity of .02058. Inclination of the orbit is 1.5610 degrees. The orbital period is 364.97 days. It rotates in 2.75 minutes.

This member of the Aten group has an albedo of about .1. The orbit allows Earth to catch it into a temporary orbital loop around Earth every 28 years.

2010 TK7 was discovered October 2010, by WISE, with Paul Wiegert finding it in the satellite's data. The diameter is about 1,000 feet.

The asteroid has an aphelion of 1.1909 AU, a perihelion of 0.8094 AU, and an eccentricity of .1907. The inclination of the orbit is 20.885 degrees. The period is 365.33 days. It is the first ever asteroid found to be moving as a Trojan of Earth, at the L4 (leading, or Greek) position.

The albedo is about .1, with an apparent magnitude that runs from 20.86 to 23.6 as it travels in a giant loop around the L4 point. Distance from the Earth ranges from as close as 12.4 million miles to over 150 million miles. (Remember that the L4 or L5 position each forms an equilateral triangle with the Sun and the planet, so for Earth it will always *average* about 93-million miles away.)

2011 QF99 was discovered in August 2011, by Michael Alexandersen and colleagues. It has a diameter of about 36 miles.

The aphelion is 22.553 AU, the perihelion is 15.772 AU, and eccentricity is .1769. This is the first discovered Trojan for the planet Uranus in its L4 position. This orbit is not stable on very long periods, and will eventually be ejected into an independent solar orbit. The inclination (which is part of the reason for the foregoing) is 10.80 degrees. The orbital period is 30638.82 days, or 83.88 years.

The albedo is believed to be about .05.

2012 XE133 was discovered December 12, 2012, by J. A. Johnson. It is about 290 feet across.

The aphelion is 1.0361 AU, the perihelion is 0.4098 AU, and the eccentricity is .4332. The orbit is inclined 6.711 degrees to the ecliptic. The orbital period is 224.53 days, or 0.615 year. The orbit shuttles between Venus's L3 and L5 positions and is regarded as a Venus co-orbital, although it also crosses Earth's orbit.

2013 LR6 was discovered April 18, 2013. It has a diameter of 38 feet.

The aphelion is 2.9264 AU, the perihelion 0.9576 AU, and the eccentricity .5069. The orbital inclination is 3.6294 degrees. The orbital period is 988.49 days, or 2.71 years. This Apollo group asteroid passed so close to Earth that it was inside the Moon's orbit in June 2013.

2013 MZ5 was discovered June 18, 2013, by Pan STARRS-1. It diameter is 1,000 feet.

The aphelion is 1.8195 AU, the perihelion is 1.2273, and the eccentricity .1751. The inclination of the orbit is 29.025 degrees. The orbital period is 703.77 days, or 1.93 years.

This was the ten thousandth Near Earth Object to be discovered and is in the Amor group.

2012VP113 discovered by Chadwick Trujillo and Scott Shepherd sets a new distance record, with an aphelion of 452 AU and a perihelion of 80 AU. It has a diameter of 280 miles.

NAMES

All names appearing in the above list are organized alphabetically here, with the number of the associated asteroid(s). For asteroids thus far nameless, their current catalog designation is given in italics.

M

N

O

P

Q

R

OBJECTS That Cross Planetary Orbits

Mercury	1566, 2002VE68
Venus	1566, 1862, 2340, 3753
Earth	1566, 1620, 1682, 1862, 2062, 2340, 3753, 4179, 4660, 6489, 25143, 54509, 99942, 101955, 162173, 163249, 164207, 2002VE68, 2006RH120, 2012XE133, 2013LR6
Mars	433, 887, 1134, 1566, 1620, 1685, 1862, 1951, 2000, 2044, 3267, 3753, 3908, 4179, 4435, 4660, 6489, 9969, 25143, 54509, 162173, 285263, 2013LR6, 2013MZ5
Jupiter	944, 5335
Saturn	944, 5145, 5335
Uranus	5145, 5335, 7066
Neptune	5145, 7066, 28978, 42355, 134340, 2001QR322

OBJECTS With Moons

2?, 18?, 22, 45, 65?, 87, 90, 121, 130, 216, 243, 624, 1089, 1313, 1862, 2044, 3673, 3749, 4492, 42355, 50000, 58534, 90482, 134340, 136108, 136199, 285263

SPECTRAL Classes

A	1951, 2732
B	2, 5, 17, 18, 729, 792, 1609, 1795, 1932, 1998, 3908
C	10, 24, 35, 50, 70, 90, 120, 121, 145, 150, 153, 156, 160, 253, 260, 275, 300, 435, 668, 998, 1691?, 2952
D	588, 944, 1578, 2241
E	44, 434, 2867
F	45, 3123
G	13, 19, 130
K	281
M	16, 22, 75, 110, 216, 250, 121514?
P	140, 190, 617, 1001, 1911, 121514?
Q	1862, 9969
S	3, 6, 7, 9, 11, 12, 14, 15, 20, 23, 25, 30, 40, 42, 60, 80, 100, 108, 158, 170, 180, 237, 243, 323, 400, 453, 480, 500, 550, 600, 800, 827, 832, 887, 951, 1000?, 1036, 1069, 1134, 1215, 1450?, 1620, 1635, 1685, 2062, 2074, 2308, 2709, 3749, 4072, 4179, 4372, 4435, 4923, 5261, 5335?, 5535, 7211
T	225, 499, 1111, 1129, 1691?, 1700?, 25143
X	87

OTHER BOOKS BY THIS AUTHOR

WORKS OF ASTRONOMY

Useful Star Names (2011)
Our Neighbor Stars (2012)
Moons of the Solar System (2013)

SCIENCE FICTION AND FANTASY

Time for Patriots (hardcover 2008, softcover 2011)--novel
The Mountain of Long Eyes (2013)--anthology of 28 stories

CPSIA information can be obtained at www.ICGtesting.com
Printed in the USA
BVOW02s1540010514

352011BV00005B/9/P